Studies in Computational Intelligence 421

Editor-in-Chief

Prof. Janusz Kacprzyk
Systems Research Institute
Polish Academy of Sciences
ul. Newelska 6
01-447 Warsaw
Poland
E-mail: kacprzyk@ibspan.waw.pl

For further volumes:
http://www.springer.com/series/7092

Sundaram Suresh, Narasimhan Sundararajan,
and Ramasamy Savitha

Supervised Learning with Complex-valued Neural Networks

Springer

Authors
Sundaram Suresh
Nanyang Technological University
Singapore

Ramasamy Savitha
Nanyang Technological University
Singapore

Narasimhan Sundararajan
Nanyang Technological University
Singapore

ISSN 1860-949X e-ISSN 1860-9503
ISBN 978-3-642-29490-7 e-ISBN 978-3-642-29491-4
DOI 10.1007/978-3-642-29491-4
Springer Heidelberg New York Dordrecht London

Library of Congress Control Number: 2012935247

Printed on acid-free paper

Springer is part of Springer Science+Business Media (www.springer.com)

Preface

Recent developments in complex-valued feed-forward neural networks have found a number of applications like adaptive array signal processing, medical imaging, communication engineering, etc. However, these applications demand an accurate phase approximation in addition to accurate magnitude approximation, which has not been well addressed in the existing literature. To fill this gap, this book focuses on the development of novel fully complex-valued feed-forward neural networks and their supervised batch and sequential learning algorithms with an emphasis on better phase approximation.

The classical approach to handle complex-valued signals is to split each complex-valued signal into two real-valued signals, either the real/imaginary components or magnitude/phase components, and then use existing real-valued neural networks. In such a split complex-valued network, real-valued activation functions and real valued weights are used to estimate the network parameters. Thus, the gradients used in updating the free parameters of the network do not represent the true complex-valued gradients resulting in poor approximation of complex-valued functions, especially the phase of the complex-valued signals. This clearly shows a need for developing fully complex-valued neural networks and their learning algorithms.

The critical issue in a fully complex-valued neural network is the selection of an appropriate activation function, as Liouville's theorem suggests that an entire and bounded function is a constant in Complex domain. However, a constant is not acceptable as an activation function as it cannot project the input space to a non-linear higher dimensional space. Also, neither analyticity nor boundedness of the activation function can be compromised, owing to the need for a stable operation of a neural network. To overcome these difficulties, Kim and Adali proposed Elementary Transcendental Functions (ETF's), which are analytic and bounded *almost everywhere (a. e)*, as activation functions for Fully Complex-valued Multi-Layer Perceptron (FC-MLP) networks and a complex-valued back-propagation learning algorithm. However, the singular points of these functions and their derivatives may interfere with the operating region of the network, thereby, affecting the performance of the network. On further investigation, we also found that the mean squared error function used to derive the gradients is only indicative of the magnitude and does not explicitly indicate the phase of the complex-valued error signal.

To overcome these difficulties, in this book, we propose an Improved Complex-valued MLP (IC-MLP) algorithm using an "*exponential*" activation function in the output layer, any ETF's as an activation function in the hidden layer and a logarithmic error function to derive the gradients. The "*exponential*" activation function ensures more stable operation of the network, as both the activation function and its derivative have their singularities at $\pm\infty$. The logarithmic error function ensures better phase approximation, as it explicitly represents both the magnitude and phase error of the function to be approximated.

Due to their localization properties, radial basis function networks exhibit good function approximation characteristics and these have been shown in real-valued applications. Therefore, we next investigate the existing complex-valued radial basis function networks in the literature. Existing complex-valued radial basis function networks use the Gaussian function as an activation function. The Gaussian function maps $\mathbb{C}^m \to \mathbb{R}$, and hence, projects the inputs from complex-valued space to a higher dimensional real-valued space in the hidden layer. Here, the centers of Gaussian neurons and output weights are complex numbers. The width of the neurons is a real number. Since the Gaussian function maps complex-valued signal to real-valued signal, real part of the error is used to update the real part of the free parameters and the imaginary part of the error is used to update the imaginary part of the free parameters. Thus, the correlation between the real and imaginary components is lost, and hence, this projection results in inaccurate phase approximation. This issue arises due to the absence of a fully complex-valued symmetric activation function.

In this book, we present a new Fully Complex-valued RBF (FC-RBF) network with a fully complex-valued activation function and its supervised learning algorithm. The fully complex-valued activation function uses the "*sech*" function as its basis function, which is symmetric about the real axis. This function maps $\mathbb{C}^m \to \mathbb{C}$ and the learning algorithm uses the fully complex-valued gradient. Existing batch learning algorithms are based on the assumption of a uniform distribution of training samples in the input space. This is not always true, especially in real-world problems. Thus, to ensure bettergeneralization ability of the network, an algorithm to select samples to participate in learning, so as to avoid over-training with similar samples, becomes essential. In this book, we develop a new self-regulatory learning system for the FC-RBF network to select samples to participate in learning in each epoch. The self-regulatory learning system chooses samples in each epoch that add significant information to the network. It also deletes samples that contain information similar to that already acquired by the network. Thus, the algorithm selects samples to be learnt or deleted in each epoch. Besides these, samples that are neither learnt nor deleted are just skipped in the present epoch, to be learnt/deleted in future epochs. The control parameters of the learning systems are also adapted based on the residual error (self-regulatory) and are independent of the problem considered.

FC-RBF and Mc-FCRBF are batch learning algorithms that use a gradient descent based learning to estiamate the free parameters of the network. Moreover, learning in these networks is based on the mean-squared error function that is only an explicit representation of the magnitude of error. While the former increases the computational complexity of the learning process, the latter may compromise on the phase approximating abilities of these networks. To address these issues, in this book, we present a novel Fully Complex-valued Relaxation Network (FCRN) with its projection based learning algorithm. FCRN is a single hidden layer network with a Gaussian-like *sech* activation function in the hidden layer and an *exponential* activation function in the output layer. The output weights are estimated by minimizing a nonlinear logarithmic function (called as an energy function) that contains an explicit representation of both the magnitude and phase error components. The projection based learning algorithm computes the optimal output weights corresponding to the minimum energy point of the energy function for a given training data set, number of hidden neurons and input weights by converting the solving of a nonlinear programming problem into that of solving a set of linear simultaneous equations. The resultant FCRN approximates the desired output more accurately with a minimal computational effort.

The performances of the proposed learning algorithms are studied in detail on two synthetic complex-valued function approximation problems, namely, Complex-valued Function Approximation Problem I (CFAP-I) and Complex-valued Function Approximation Problem II (CFAP-II). In these problems, the performances are compared with existing complex-valued learning algorithms, FC-MLP, Complex-valued Radial Basis Function network (CRBF), Complex-valued Extreme Learning Machines (CELM) and Complex-valued Minimal Resource Allocation Network (CMRAN). Next, the proposed algorithms are also evaluated in comparison to existing algorithms in the literature using two real-world problems, namely, complex-valued Quadrature Amplitude Modulation (QAM) equalization problem and an adaptive beam-forming problem. These problems involve complex-valued signals and require accurate phase approximation. It is observed that while the proposed algorithms, viz., the IC-MLP, the FC-RBF and the CSRAN are good in performing QAM equalization with lesser symbol error rates, the FC-MLP, the IC-MLP and the FC-RBF with the self-regulatory learning system perform better for beam-forming with better isolation of nulls and desired signals. Although a strict proof for stability and convergence of all the proposed algorithms are still under study, simulation studies conducted on a number of problems indicate that the proposed algorithms are more efficient (in accurate phase approximations with a compact structure) than other existing complex-valued neural network learning algorithms.

Recently, it has been proved in the literature that the orthogonal decision boundaries of the complex-valued neural networks provide them with an exceptional ability to perform real-valued classification tasks. The Multi-layer network using Multi-valued Neurons (MLMVN) and the single layer complex-valued neural network, referred to as, Phase Encoded Complex-valued Neural Network (PE-CVNN) are the two complex-valued classifiers available in the literature. The transformations used to convert the real-valued features to the Complex domain in these classifiers, do not completely exploit the advantages of orthogonal decision boundaries or are not unique (do not perform one-to-one mapping of feature from Real to Complex domain). Moreover, these algorithms iteratively update the parameters of the classifier and hence, suffer from slow convergence. Hence there is a need for a fast learning fully complex-valued classifier and a transformation that uniquely maps (one-to-one mapping) the real-valued input features to the four quadrants of the Complex domain to fully exploit the advantages of the orthogonal decision boundaries.

In this book, we present the circular complex-valued extreme learning machine classifier for performing real-value classification tasks in the Complex domain. The CC-ELM classifier uses a single hidden layer network with a non-linear input/hidden layer and a linear output layer. At the input layer, a unique nonlinear circular transformation is used as the activation function to make a one-to-one mapping of the real-valued input features to the Complex domain. At the hidden layer, the complex-valued input features are mapped on to a higher dimensional Complex plane, using the '*sech*' activation function. In CC-ELM, the input parameters and the parameters of the hidden layer are chosen randomly and the output weights are calculated analytically, requiring lesser computational effort to perform classification tasks. Existence of orthogonal decision boundaries in the hidden and output layers of the CC-ELM has been proven analytically. The advantages of these orthogonal decision boundaries and the activation function used in the input layer are studied by comparing the performance of the CC-ELM classifier with other real-valued and complex-valued classifiers using benchmark classification problems from the UCI machine learning repository and two practical classification problems. Performance of CC-ELM on these data sets clearly shows that it performs better than the existing (well-known) real-valued and other complex-valued classifiers, especially for unbalanced data sets.

Complex-valued sequential learning algorithms have been developed to handle a large training data set or when the data is subject to temporal changes or when the complete training dataset is not available a priori. In a sequential learning algorithm, samples are presented one-by-one, only once and are deleted after being learnt. In such a sequential mode of learning, the network can be made to grow and prune itself, until it achieves a compact structure. Thus, it does not require the network size to be fixed a priori. However, complex-valued sequential learning algorithms, available in the literature, use the Gaussian function that maps $\mathbb{C}^m \to \mathbb{R}$, as the activation function. Hence, they suffer from drawbacks similar to that of the CRBF networks stated above.

In this book, we propose a new Complex-valued Self-regulatory Resource Allocation Network (CSRAN), using the above self-regulatory learning scheme in a sequential framework. CSRAN alters the training sequence, and evolves the network to achieve compact structure with a desired performance. The basic building block of CSRAN is the FC-RBF structure. In a sequential framework, the self-regulatory learning system, deletes the sample, learns the samples (add neuron or update the parameters) or shifts the sample to the rear end of the data stream for future use. Similar to the batch learning framework, the control parameters are also self-regulatory. The influence of the various control parameters on the performance of the CSRAN is studied. Based on this study, we present a guideline on the initialization of the control parameters of the CSRAN. The performance study on a set of function approximation and classification problems clearly indicates the advantage of the CSRAN over other complex-valued learning algorithms existing in the literature.

Contents

1 Introduction 1
1.1 Nature of Complex-valued Neural Networks 2
1.1.1 Split Complex-valued Neural Network 2
1.1.2 Fully Complex-valued Neural Networks 4
1.2 Types of Learning 8
1.2.1 Supervised Learning 8
1.2.2 Unsupervised Learning 13
1.3 Mode of Learning 17
1.3.1 Complex-valued Batch Learning Algorithms 17
1.3.2 Complex-valued Sequential Learning Algorithms 19
1.4 Applications 20
1.4.1 Digital Communication: QAM Equalization 21
1.4.2 Array Signal Processing 21
1.4.3 Real-Valued Classification 22
1.4.4 Memories 23
1.4.5 Other Applications 23
References 24

2 Fully Complex-valued Multi Layer Perceptron Networks 31
2.1 Complex-valued Multi-Layer Perceptron Networks 32
2.1.1 Split Complex-valued Multi-Layer Perceptron 32
2.1.2 Fully Complex-valued Multi-Layer Perceptron 35
2.2 Issues in Fully Complex-valued Multi-Layer Perceptron Networks 40
2.2.1 Split Complex-valued MLP 41
2.2.2 Fully Complex-valued MLP 41
2.3 An Improved Fully Complex-valued Multi-Layer Perceptron (IC-MLP) 42
2.3.1 A New Activation Function: *exp* 43
2.3.2 Logarithmic Performance Index 45
2.3.3 Learning Algorithm 46

2.4 Summary 46
References 47

3 A Fully Complex-valued Radial Basis Function Network and Its Learning Algorithm 49
3.1 Complex-valued RBF Networks 50
3.2 Factors Influencing the Performance of Complex-valued RBF Networks 53
3.3 A Fully Complex-valued RBF Network (FC-RBF) 54
3.3.1 Network Architecture 54
3.3.2 The Activation Function 54
3.4 Learning Algorithm for the FC-RBF Network 56
3.4.1 Network Initialization: K-means Clustering Algorithm 60
3.5 Meta-cognitive Fully Complex-valued Radial Basis Function Network 61
3.5.1 Cognitive Component of Mc-FCRBF: The FC-RBF Network 63
3.5.2 Meta-cognitive Component of Mc-FCRBF: Self-regulatory Learning Mechanism 63
3.6 Summary 69
References 70

4 Fully Complex-valued Relaxation Networks 73
4.1 Fully Complex-valued Relaxation Networks 74
4.1.1 FCRN Architecture 74
4.1.2 Nonlinear Logarithmic Energy Function 76
4.1.3 A Projection Based Learning Algorithm for FCRN 77
4.2 Summary 82
References 83

5 Performance Study on Complex-valued Function Approximation Problems 85
5.1 Synthetic Function Approximation Problems 85
5.1.1 Synthetic Complex-valued Function Approximation Problem I (CFAP-I) 86
5.1.2 Synthetic Complex-valued Function Approximation Problem II (CFAP-II) 88
5.2 Real-World Problems 89
5.2.1 Complex Quadrature Amplitude Modulation Channel Equalization Problem 89
5.2.2 Cha and Kassam Channel Model 92
5.2.3 Adaptive Beam-Forming Problem 96
5.3 Summary 105
References 106

6 Circular Complex-valued Extreme Learning Machine Classifier 109
6.1 Complex-valued Classifiers in the Literature 110
6.1.1 Description of a Real-valued Classification Problem Done in the Complex Domain 110
6.1.2 Multi-Layer Neural Network Based on Multi-Valued Neurons (MLMVN) 111
6.1.3 Phase Encoded Complex-Valued Neural Network (PE-CVNN) 112
6.1.4 Modifications in FC-MLP, FC-RBF and Mc-FCRBF Learning Algorithm to Solve Real-valued Classification Problems 113
6.2 Circular Complex-valued Extreme Learning Machine Classifier ... 114
6.2.1 Architecture of the Classifier 114
6.2.2 Learning Algorithm of CC-ELM 117
6.2.3 Orthogonal Decision Boundaries in CC-ELM 118
6.2.4 Case (i): Orthogonality of Decision Boundaries in the Output Layer 118
6.2.5 Case (ii): Orthogonality of Decision Boundaries in the Hidden Layer 120
6.3 Summary 122
References 123

7 Performance Study on Real-valued Classification Problems 125
7.1 Descriptions of Real-valued Benchmark Classification Problems 125
7.2 Performance Study 126
7.2.1 Performance Measures 126
7.2.2 Multi-category Real-valued Classification Problems 127
7.2.3 Binary Real-valued Classification Problems 129
7.3 Performance Study Using a Real-world Acoustic Emission Classification Problem 129
7.4 Summary 132
References 132

8 Complex-valued Self-regulatory Resource Allocation Network (CSRAN) 135
8.1 A Brief Review of Existing Complex-valued Sequential Learning Algorithms 137
8.2 Complex-valued Minimal Resource Allocation Network (CMRAN) 138
8.2.1 Drawbacks of the CMRAN Algorithm 143
8.3 Complex-valued Growing and Pruning RBF (CGAP-RBF) Networks 143
8.4 Incremental Fully Complex-valued Extreme Learning Machines (I-ELM) 146

8.5 Complex-valued Self-regulatory Resource Allocation Network Learning Algorithm (CSRAN) . . . 146
8.5.1 Network Architecture . . . 146
8.5.2 Sequential Self-regulating Learning Scheme of CSRAN . . . 148
8.5.3 Guidelines for the Selection of the Self-regulatory Thresholds . . . 152
8.5.4 Illustration of the Self-regulatory Learning Principles Using a Complex-valued Function Approximation Problem . . . 156
8.6 Performance Study: Complex-valued Function Approximation Problems . . . 160
8.6.1 Complex-valued Function Approximation Problem I . . . 160
8.6.2 Complex-valued Function Approximation Problem II . . . 160
8.6.3 QAM Channel Equalization Problem . . . 162
8.6.4 Adaptive Beam Forming Problem . . . 163
8.7 Performance Study: Real-valued ClassificationProblems . . . 165
8.8 Summary . . . 166
References . . . 167

List of Acronyms

a.e	*almost everywhere*.
CBP	Complex-valued Back Propagation.
CC-ELM	Circular Complex-valued Extreme Learning Machine.
CEKF	Complex-valued Extended Kalman Filter
C-ELM	Complex-valued Extreme Learning Machines.
CFAP-I	Complex-valued Function Approximation Problem I.
CFAP-II	Complex-valued Function Approximation Problem II.
CFAPs	Complex-valued Function Approximation Problems.
CGAP-RBF	Complex-valued Growing and Pruning Radial Basis Function Network.
CICA	Complex-valued Independent Component Analysis.
CMLP	Complex-valued Multi-layered Perceptron.
CMRAN	Complex-valued Minimal Resource Allocation Network.

CRBF	Complex-valued Radial Basis Function Network.
CSRAN	Complex-valued Self-regulatory Resource Allocation Network.
CVNNs	Complex-valued Neural Networks.
CXOR	Complex-valued XOR.
EKF	Extended Kalman Filter.
ETFs	Elementary Transcendental Functions.
FC-MLP	Fully Complex-valued Multi-layer Perceptron.
FCNNs	Fully Complex-valued Neural Networks.
FC-RBF	Fully Complex-valued Radial Basis Function.
FCRN	Fully Complex-valued Relaxation Network
GAP-RBF	Growing and Pruning Radial Basis Function Network.
IC-MLP	Improved Complex-valued Multi-layer Perceptron.
I-ELM	Incremental Extreme Learning Machines.
SMC-RBF	Sequential Multi Category Radial Basis Function
MLMVN	Multi Layered Multi Valued Network
PE-CVNN	Phase Encoded Complex Valued Neural Network
i.i.d	*independent and identically distributed.*
KMC	K-Means Clustering.
MLP	Multi-layer Perceptron.
MLMVN	Multi-layered Multi-valued Neural Network
MRAN	Minimal Resource Allocation Network.

NLCPCA	Non-linear Complex-valued Principal Component Analysis.
PCA	Principal Component Analysis.
PE-CVNN	Phase Encoded Complex-valued Neural Network
QAM	Quadrature Amplitude Modulation.
RBF	Radial Basis Function.
RMS	Root Mean Squared.
SC-MLP	Split Complex-valued Multi-layer Perceptron.
SER	Symbol Error Rate.
w.r.t	*with respect to*.

List of Symbols

Scalars:

a, p, x, u	Real part of a complex-valued signal
b, q, y, v	Imaginary part of a complex-valued signal
c_1, c_2, c_3, c_4	Function variables (of the function to be approximated)
d	Distance between antenna elements
E_a^M	Magnitude control parameter for addition
E_a^ϕ	Phase control parameter for addition
E_d^M	Magnitude control parameter for deletion
E_d^ϕ	Phase control parameter for deletion
E_l^M	Magnitude control parameter for learning
E_l^ϕ	Phase control parameter for learning
E_P	Pruning threshold
f	Frequency of the beam
F	Percentage of failure
h	Number of hidden neurons in a fully complex-valued neural network
i	$\sqrt{(-1)}$, the Complex operator
J_{Me}	Root mean squared magnitude error
m	Number of input neurons in a fully complex-valued neural network

M_t^e Instantaneous magnitude error of sample

n Number of output neurons in a fully complex-valued neural network

N Number of samples in a training dataset

N_{all} Number of experiments at a given learning rate in the convergence study

N_D Number of Deleted samples

N_f Number of Feedback connections (Recurrent networks)

N_L Number of samples participated in learning

N_S Number of samples skipped from learning

N_U Number of samples used in training (addition and updation of network parameters)

r_{hRe}^k Normalized real component of hidden neuron output

r_{hIm}^k Normalized imaginary component of hidden neuron output

$s(n)$ Transmitted symbols (Channel equalizer)

s_i^1 Sensitivity of a single neuron in the hidden layer

s_i^2 Sensitivity of a single neuron in the output layer

v_{kj}^0 Complex-valued weight connecting the jth input neuron and the kth hidden neuron

v_{kj}^1 Complex-valued weight connecting the jth hidden neuron and the kth output neuron

w_{kj}^0 Real-valued weight connecting the jth input neuron and the kth hidden neuron

w_{kj}^1 Real-valued weight connecting the jth hidden neuron and the kth output neuron

y_h^i Output of the ith hidden neuron of the split complex-valued neural network

$y_q(a)$ States of the channel symbol

z_h^i Output of the ith hidden neuron of the fully complex-valued network

Vectors:

$\mathbf{c}_j$	Center of the j-th hidden neuron (RBF Network)
$\mathbf{g}$	Vector of the complex-valued variables (function approximation problems)
$\mathbf{o}$	Real-valued target to the split complex-valued neural network
$\widehat{\mathbf{o}}$	Real-valued output of a split complex-valued neural network
$\mathbf{v}_j^0$	Complex-valued input weight vector connecting the jth hidden neuron of the fully complex-valued neural network
$\mathbf{s}^1$	Sensitivity of the hidden layer
$\mathbf{s}^2$	Sensitivity of the output layer
$\mathbf{v}_j^1$	Complex-valued output weight vector connecting the jth hidden neuron of the fully complex-valued neural network
$\mathbf{w}_j^0$	Real-valued input weight vector connecting the jth hidden neuron of the split complex-valued neural network
$\mathbf{w}_j^1$	Real-valued output weight vector connecting the jth hidden neuron of the split complex-valued neural network
$\mathbf{y}$	complex-valued target vector to the fully complex-valued neural network
$\widehat{\mathbf{y}}$	Complex-valued output of a fully complex-valued neural network
$\mathbf{y}_q(n)$	Channel observation vector
$\mathbf{z}$	Complex-valued input vector to the neural network
$\mathbf{z}_R$	Real component of the complex-valued input vector signal $\mathbf{z}$
$\mathbf{z}_I$	Imaginary component of the complex-valued input vector signal $\mathbf{z}$

Matrices and Functions:

V^0	Complex-valued input weight matrix connecting input layer and hidden layer of the fully complex-valued neural network; $V^0 \in \mathbb{C}^{h \times m}$
V^1	Complex-valued output weight matrix connecting hidden layer and output layer of the fully complex-valued neural network;$V^1 \in \mathbb{C}^{n \times h}$
W^0	Real-valued input weight matrix connecting input layer and hidden layer of the split complex-valued neural network; $W^0 \in \mathbb{R}^{h \times 2m}$
W^1	Real-valued output weight matrix connecting hidden layer and output layer of the split complex-valued neural network; $W^1 \in \mathbb{R}^{2n \times h}$
$\sigma(.)$	Real-valued activation function $\mathbb{R}^{2m} \to \mathbb{R}$
$f_a(\cdot)$	Complex-valued activation function $\mathbb{C}^m \to \mathbb{C}$
$f_{aR}(\cdot)$	Real part of the axial symmetric activation function
$f_{aI}(\cdot)$	Imaginary part of the axial symmetric activation function
$f_g(\cdot)$	Gaussian activation function in complex-valued radial basis function networks $f_g(.) : \mathbb{C}^m \to \mathbb{R}$
$E(.)$	Mean squared error function
$E_{new}(.)$	Logarithmic error function

Greek Symbols:

α	Constant used in axial symmetric activation function
δ	Slope of control parameters
δ_j	Complex-valued error gradient
δ_n	Time delay between the first and n^{th} antenna element (beamformer system)
η	Learning rate parameters used in gradient based learning algortihms
η_a	Average Per class classification accuracy
η_o	Overall classification accuracy
$\eta_{\mathbf{v}}$, $\eta_{\mathbf{c}}$, $\eta_{\boldsymbol{\sigma}}$	Learning rates for the free parameters of the fully complex-valued radial basis function network
λ	Wavelength of the beam
μ_v, μ_σ and μ_c	Learning rates for the free parameters of the complex-valued radial basis function network
ϕ_e	Average phase error in deg.
ϕ_t^e	Instantaneous phase error of sample t
σ_j	Gaussian width of the jth hidden neuron (real-valued scalar)
$\boldsymbol{\sigma}_j$	Hidden neuron weight of the jth hidden neuron in the FC-RBF Network (complex-valued vector)
τ	Equalizer delay
θ	Angle of arrival

Chapter 1
Introduction

Signals occurring in applications like medical imaging and telecommunications are inherently complex-valued, and processing them in their natural form preserves the physical characteristics of these signals. Therefore, there is a widespread research interest in developing efficient complex-valued neural networks along with their learning algorithms. However, operating in the Complex domain presents new challenges; foremost among them being the choice of an appropriate complex-valued activation function. Basically, an activation function for a neural network is required to be nonlinear, bounded and differentiable in every point on the considered plane [1]. This implies that in the Complex domain, the function has to be nonlinear, bounded and entire. However, Liouville's theorem states that an entire and bounded function in the Complex domain is a constant (function) [2]. As neither the analyticity and boundedness can be compromised, nor is a constant function acceptable as an activation function as it cannot project the input space to a non-linear higher dimensional space, choices for activation functions for complex-valued neural network are limited. In this chapter, the different complex-valued neural networks existing in the literature are discussed in detail, along with their limitations.

Complex-valued neural networks can be broadly classified into different categories as shown below:

- Nature of Complex-valued Neural Networks
  - Split complex-valued neural network
  - Fully complex-valued neural network
- Type of Learning
  - Supervised learning
  - Unsupervised learning
  - Reinforcement learning
- Mode of Learning
  - Batch learning
  - Sequential learning

S. Suresh et al.: Supervised Learning with Complex-valued Neural Networks, SCI 421, pp. 1–29.
springerlink.com

- Applications
  - Array signal processing
  - Wireless communication: QAM equalization
  - Memories
  - Real-valued classification and other applications

1.1 Nature of Complex-valued Neural Networks

Based on the nature of signals (i.e., the inputs signals/features, the output signals/variables and the weight parameters), which are real-valued or complex-valued, the complex-valued neural networks are classified into split complex-valued neural networks and fully complex-valued neural networks.

1.1.1 Split Complex-valued Neural Network

Initially, split complex-valued networks [3] were used to operate on complex-valued signals. The split complex-valued networks are further divided into two types based on the nature of weights, namely, split complex networks with real-valued weights and real-valued activation functions and split complex networks with complex-valued weights and real-valued activation functions.

1.1.1.1 Real-Valued Weights and Real-Valued Activation Functions

The neural network architecture is similar to the classical Multi-Layer Perceptron (MLP) network with the real-valued back propagation algorithm. Here, the complex-valued inputs and targets are split into two real-valued quantities (the real and imaginary components), based on either a rectangular (real-imaginary) or polar (magnitude-phase) coordinate system. For example, two complex-valued inputs and one complex-valued output problem is converted into four real-valued inputs and two real-valued outputs problem such as:

$$\begin{Bmatrix} z_1 \\ z_2 \end{Bmatrix} \rightarrow \begin{Bmatrix} z_{1R} \\ z_{1I} \\ z_{2R} \\ z_{2I} \end{Bmatrix} \tag{1.1}$$

$$\begin{Bmatrix} z_1 \\ z_2 \end{Bmatrix} \rightarrow \begin{Bmatrix} r_1 \\ \phi_1 \\ r_2 \\ \phi_2 \end{Bmatrix} \tag{1.2}$$

where z_{1R} and z_{1I} are the real and imaginary parts of the complex-valued signal z_1 and r_1 and ϕ_1 are its magnitude and phase components, respectively. Similarly, z_{2R}, z_{2I}, r_2 and ϕ_2 are the real part, imaginary part, magnitude and phase of the complex-valued signal z_2. The input and output weights of such a network are also real-valued. The architecture of a such split complex-valued multi-layer perceptron network with a single hidden layer is shown in Fig. 1.1.

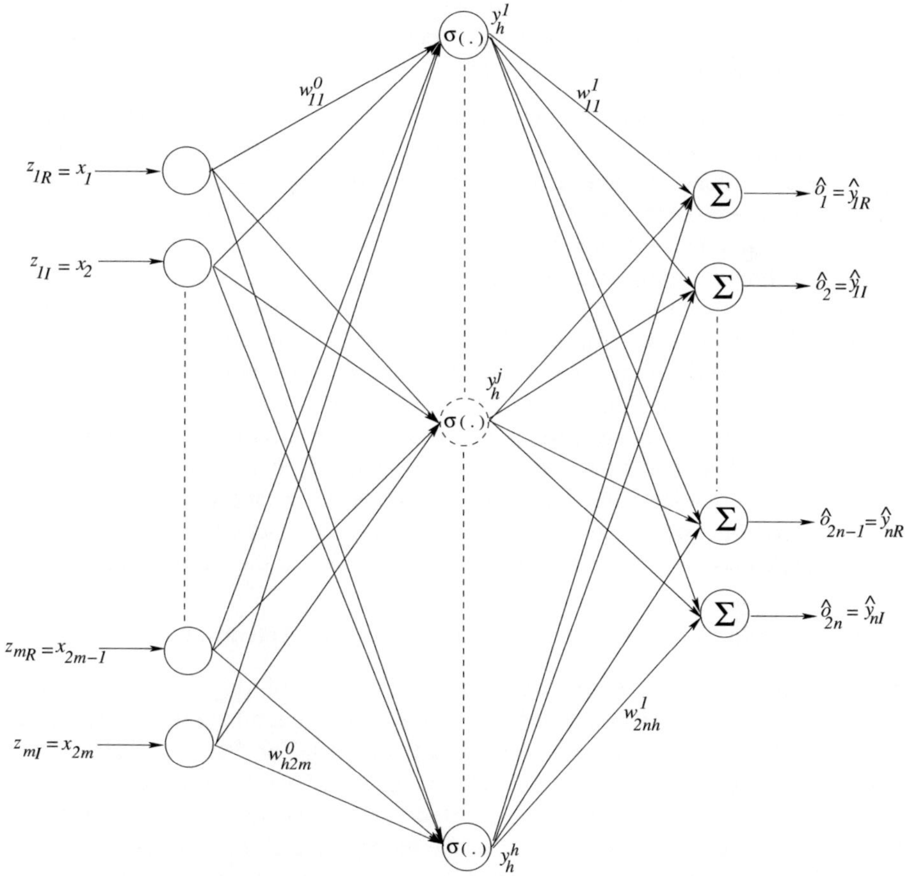

Fig. 1.1 Architecture of a Split Complex-valued Multi-Layer Perceptron Network

In the figure,

- $\mathbf{x} \in \mathbb{R}^{2m} = [x_1,\ x_2,\ \cdots,\ x_{2m}] = [z_{1R},\ z_{1I},\ \cdots\ z_{mR},\ z_{mI}]^T$ are the real-valued inputs.
- $W^0 \in \mathbb{R}^{h\times 2m} = \begin{bmatrix} w_{11}^0 & \cdots & w_{12m}^0 \\ \vdots & & \\ w_{h1}^0 & \cdots & w_{h2m}^0 \end{bmatrix}$ are the real-valued weights connecting the input layer and hidden layer.
- $\mathbf{y}_h \in \mathbb{R}^h = [y_h^1,\ y_h^2,\ \cdots,\ y_h^h]^T$ are the real-valued response of the hidden neurons given by $y_h^k = \sigma(\sum_{j=1}^{2m} w_{kj}x_j)$; $\sigma(.)$ is a unipolar or bipolar sigmoidal activation function.

- $W^1 \in \mathbb{R}^{2n \times h} = \begin{bmatrix} w^1_{11} & \cdots & w^1_{12m} \\ \vdots & & \\ w^1_{h1} & \cdots & w^1_{h2m} \end{bmatrix}$ are the real-valued weights connecting the hidden layer and output layer.
- $\widehat{\mathbf{o}} \in \mathbb{R}^{2n} = [\widehat{o}_1, \; \cdots, \; \widehat{o}_{2n}]^T$ are the real-valued outputs of the network. The complex-valued predicted output of a split complex-valued multi-layer perceptron is reconstructed as $\widehat{y}_1 = \widehat{o}_1 + i\widehat{o}_2$.

Hence, it can be observed that the network is in fact an MLP network operating on real-valued signals. As each complex-valued signal is split into two real-valued signals, the number of neurons in the input and output layers are twice the number of inputs and outputs (of the function to be approximated). This results in an increased network size and hence, more parameters to be estimated. Also, irrespective of the kind of splitting the complex-valued signal, real-valued gradients are used to update the network parameters of such a split complex-valued network. These real-valued gradients do not reflect the true complex-valued gradient, as stated by Kim and Adali [4]. Hence, approximation using the split complex-valued neural networks results in inaccurate approximations and introduces phase distortion in the complex-valued function that is approximated. Moreover, from the sensitivity analysis on the split complex-valued networks presented by Yang *et al.* [5] and from the convergence study on split complex-valued networks presented by Zhang *et al.* [6], it can be observed that the convergence of split complex-valued network with real-valued weights depends on proper initialization and the choice of the learning rate [7].

1.1.1.2 Complex-Valued Weights and Real-Valued Activation Functions

To overcome the problem of phase distortions due to the splitting of complex-valued signal, complex-valued weights with real-valued activation functions have been presented by Deng *et al.* [8, 9] and Benvenuto *et al.*, [10]. The Gaussian function that maps the Complex domain into the Real domain, $f : \mathbb{C}^n \rightarrow \mathbb{R}$ is used as the basis of the activation function in these networks. Hence, these networks also use real-valued activation functions and the response of the hidden layer is real-valued. In [8, 10], the product of complex-valued error and its conjugate is used to update the weights during error-correction learning. As the Real part of error is used to update the Real part of the free parameters, and Imaginary part of error to update Imaginary part of the free parameters, the correlation between the real and imaginary components of the weight and error is lost during learning. Hence the gradients, used to update the parameters during learning are not true representation of the complex-valued gradients [4]. Therefore, approximation using such a network is also not very accurate.

1.1.2 Fully Complex-valued Neural Networks

The fully complex-valued network with a fully complex-valued activation function is capable of handling complex-valued inputs and outputs efficiently. The architecture of a fully complex-valued neural network is given in Fig. 1.2.

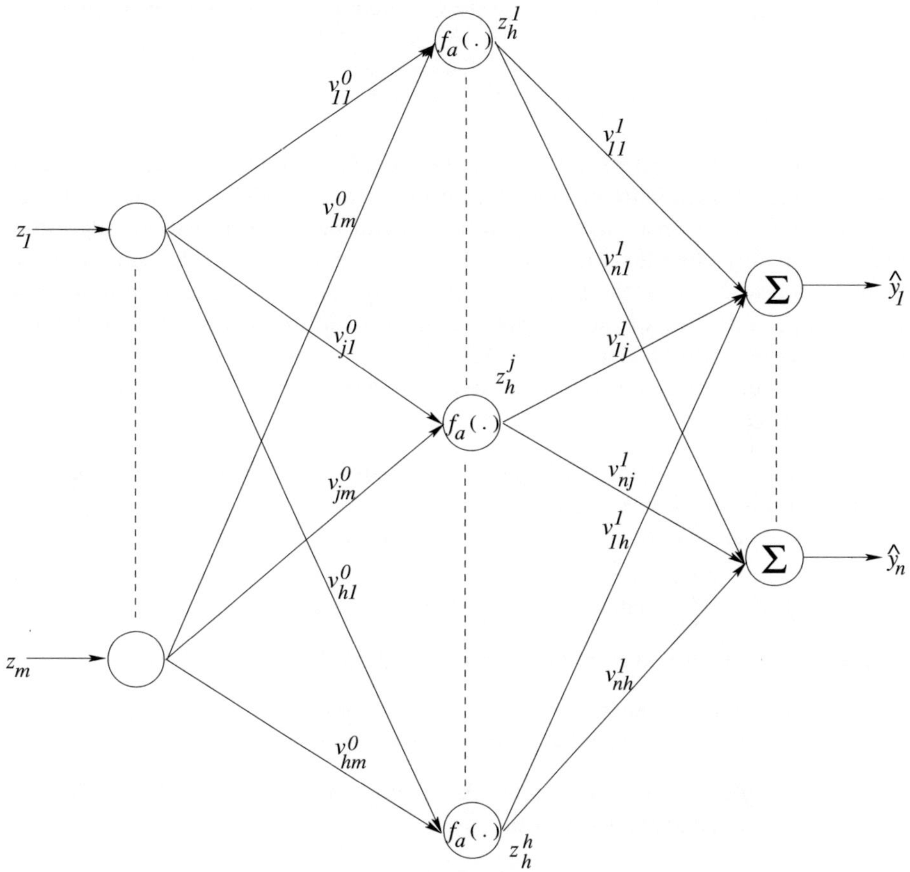

Fig. 1.2 Architecture of a Fully Complex-valued Neural Network

In the figure,

- $\mathbf{z} \in \mathbb{C}^m = [z_1,\ z_2,\ \cdots,\ z_m]^T$ are the complex-valued inputs to the network.
- $V^0 \in \mathbb{C}^{h\times m} = \begin{bmatrix} v_{11}^0 & \cdots & v_{1m}^0 \\ \vdots & & \\ v_{h1}^0 & \cdots & v_{hm}^0 \end{bmatrix}$ are the complex-valued weights connecting the input layer and hidden layer.
- $\mathbf{z}_h \in \mathbb{C}^h = [z_h^1,\ z_h^2,\ \cdots,\ z_h^h]^T$ are the complex-valued response of the hidden neurons given by $y_h^k = f_a(\sum_{j=1}^m v_{kj}^0 z_j;\ k = 1,2,\cdots h)$. Here, $f_a(.)$ is a complex-valued activation function with sigmoidal characteristics.
- $V^1 \in \mathbb{C}^{n\times h} = \begin{bmatrix} v_{11}^1 & \cdots & v_{1m}^1 \\ \vdots & & \\ w_{h1}^1 & \cdots & w_{hm}^1 \end{bmatrix}$ are the complex-valued weights connecting the hidden layer and output layer.

- $\widehat{\mathbf{y}} \in \mathbb{C}^n = [\widehat{y}_1, \cdots, \widehat{y}_n]^T$ are the complex-valued outputs of the network given by

$$\widehat{y}_l = \sum_{k=1}^{K} v_{lk}^1 z_h^k \tag{1.3}$$

It can be observed from the figure that unlike the split complex-valued networks, the fully complex-valued networks operate on the complex-valued signals, and hence use only a fewer neurons in the input and output layers. The learning algorithm used in the fully complex-valued neural network relies on well-defined complex-valued gradients[1]. Thus, the main issue in a fully complex-valued neural network is the proper selection of the complex-valued activation function and the computation of its derivatives. For the activation function selection, the complex-valued function should satisfy the following essential properties stated by Georgiou and Koutsougeras in [12] :

- $f_a(z) = f_a(x+iy) = u(x,y) + iv(x,y)$. $u(x,y)$ and $v(x,y)$ should be non-linear and bounded in x and y.
- The partial derivatives $\frac{\partial u}{\partial x}, \frac{\partial u}{\partial y}, \frac{\partial v}{\partial x}, \frac{\partial v}{\partial y}$ exist and are bounded.
- $f_a(z)$ is not entire[2].
- $\frac{\partial u}{\partial x}\frac{\partial v}{\partial y} \neq \frac{\partial v}{\partial x}\frac{\partial u}{\partial y}$ unless $\frac{\partial u}{\partial x} = \frac{\partial v}{\partial x} = 0$ and $\frac{\partial v}{\partial y} = \frac{\partial u}{\partial y} = 0$

These conditions were, then reduced and relaxed by Kim and Adali [4] as:

> *In a bounded domain of complex plane $\mathbb{C}$, a fully complex nonlinear activation function $f_a(z)$ needs to be analytic[3] and bounded almost everywhere.*

The fully complex-valued neural network is classified further based on the nature of the complex-valued activation function used.

1.1.2.1 Elementary Transcendental Activation Functions

Kim and Adali introduced a Fully Complex-valued MLP neural network with any of the Elementary Transcendental Functions (ETF's) as an activation function and derived its fully complex-valued back propagation weight update rule [4]. The learning algorithm presented in [4] is a complex-valued version of the real-valued back-propagation algorithm. For complete details on the properties of different ETF's and the complex-valued back propagation algorithm, refer to [4]. The ETF's satisfy the properties needed for complex activation functions, but they along with their derivatives have essential, removable or isolated singularities at different locations in the complex domain [13]. For example, *asinh* is a simple ETF, which has a branch cut singularity along the imaginary axis. The derivative of *asinh* has an isolated

[1] If $z = x+iy$ and $f_a(z) = u(x,y)+iv(x,y)$, $\frac{\partial f_a}{\partial z} = \frac{1}{2}\left(\frac{\partial f_a}{\partial x} - i\frac{\partial f_a}{\partial y}\right)$ as defined in [11].

[2] In complex analysis, an entire function, also called an integral function, is a complex-valued function that is analytic over the whole complex plane.

[3] A complex function is said to be analytic on a region $\mathbb{C}$ if it is complex differentiable at every point in $\mathbb{C}$.

Table 1.1 ETF's and their singularities

Activation Function	Singular Point	Derivative	Singular point of the derivative
$tanh$	Pointwise: $(1/2+n)pi$	$sech^2$	Pointwise: $(1/2+n)\pi$
tan	Pointwise: $(1/2+n)pi$	sec^2	Pointwise: $(1/2+n)\pi$
$atanh$	Isolated: $\pm$ 1	$1/1-z^2$	Isolated: $\pm$ 1
$atan$	Isolated: $\pm i$	$1/1+z^2$	Isolated: $\pm i$
$asinh$	Branch Cut: along imag axis $\geq$ i	$1/\sqrt(1+z^2)$	Isolated: $\pm i$
$asin$	Branch Cut: along real axis $\geq$ 1	$1/\sqrt(1-z^2)$	Isolated: $\pm$ 1

singularity at $\pm$ i. The list of various ETF's, and their derivatives, along with the singularities associated with the ETF and its derivative are presented in Table 1.1.

If the learning algorithm operates in the region of singularity, the parameter updates exhibit undesirable behavior affecting the convergence. Hence, the fully complex-valued algorithm is sensitive to the singularities of the activation functions, sample population distribution, weight initialization and the choice of the learning rate. It is also noteworthy that the convergence effects are different for different ETF's and they also depend on the nature of the problem, the initial weights and the learning rate. Hence, it is essential to identify a fully complex-valued activation function that is less sensitive to initial weights, learning rate and the the problem considered. One such activation function is proposed for FC-MLP networks in Chapter 2. Besides, there is also a need to develop a fully complex-valued activation function for a fully complex-valued radial basis function network. Such an activation function and its learning algorithm are presented in Chapter 3.

1.1.2.2 Axially Symmetric Activation Function

To overcome the difficulty of boundedness and analyticity of the complex-valued activation function, You and Hong introduced an axially symmetric complex-valued function to deal with QAM signals [14]. The general form of the axially-symmetric activation function is given by

$$f_a(z) = f_{aR}(x) + i f_{aI}(y) \tag{1.4}$$

where $z \varepsilon C$, $z_R = real(z)$, $z_I = imag(z)$ and $f_{aR}(.)andf_{aI}(.)$ are any continuous functions. You and Hong used

$$f_{aR}(x) = x + \alpha sin(\pi x); \text{ and} f_{aI}(y) = y + \alpha sin(\pi y) \tag{1.5}$$

where α is the slope factor that determines the degree of non-linearity. Here, $0 < \alpha < \frac{1}{\pi}$. The axially symmetric function satisfies the essential properties of an activation function to be used in the Complex domain. The axially symmetric function is suitable for problems, which have symmetric targets, such as equalization in telecommunication. Even though the axially symmetric activation function does not have the singularity, it does not consider the correlation between the real and imaginary parts of a complex-valued signal. Also, selection of appropriate continuous activation function for a given problem is difficult [15, 7].

1.2 Types of Learning

As in real-valued networks, complex-valued networks can also be classified into supervised learning and unsupervised learning types, depending on the presence or absence of a teacher.

1.2.1 Supervised Learning

If the learning in a neural network occurs with a teacher, it is called *Supervised Learning*. In supervised learning, the teacher has a knowledge of the environment, with the knowledge being represented by a set of input-output samples, called the training dataset [1]. The objective of learning is to estimate the free parameters of the network, such that the output errors are minimized. Several supervised learning schemes are available for complex-valued networks in the framework of feed-forward neural networks and recurrent neural networks.

1.2.1.1 Feed-forward Neural Networks

Neural networks in which signals are transmitted from input nodes to output nodes, with no feedback or memory are defined as feed-forward neural networks [1]. Two well-known topologies of complex-valued feed-forward networks are complex-valued Multi-Layer Perceptron (cMLP) networks and complex-valued Radial Basis Function (cRBF) networks.

Complex-valued MLP networks: Similar to the real-valued networks, complex-valued MLP networks are popular learning paradigms based on error correction learning. The training/learning is based on a set of inputs-outputs, and learning occurs based on the minimization of the error functions. Complex-valued Back Propagation (CBP) algorithm for the complex-valued MLP network was first presented by Leung and Haykin [3]. Georgiou and Koutsougeras [12] listed the essential properties of a fully complex-valued activation function and presented an improved version of the complex-valued back-propagation algorithm. Later, Kim and Adali [4] relaxed these conditions and presented an improved version of the CBP using the Cauchy Riemann's equations[4]. Besides these, the Fully Complex-valued Extreme Learning Machine (C-ELM) presented by Li *et al.* [16] is also a direct extension of the real-valued Extreme Learning Machine (ELM) presented by Huang *et al.* [17, 18]. C-ELM algorithm determines the free parameters of thc network in an analytical way and does learning faster compared to other algorithms. Similarly, the complex-valued resilient propagation network presented by Kantsila *et al.* [19] is a direct extension of the real-valued resilient propagation network [20].

In all these CBP algorithms, the activation functions are bounded *almost everywhere (a.e)*. This is because, according to Liouville's theorem,a bounded entire activation function in the Complex domain is a constant function . To overcome the controversy of boundedness and analyticity of complex-valued activation function, an axially symmetric complex-valued function is introduced, to deal with QAM signals, by You and Hong [14]. However, the issue here, is the selection of appropriate functions for the real and imaginary parts of the function (f_{aR} and f_{aI} (1.5)). Therefore, there is a need to identify a fully complex-valued activation function, which is bounded in the bounded domain of the Complex plane, with its singular point away from the operating region of the fully complex-valued neural network. In other words, a fully complex-valued activation function with its singularity at $\pm\infty$ is preferred. One such activation function is discussed in Chapter 2. book.

Complex-valued RBF networks: As the real-valued radial basis function networks form another popular architecture and are well-known for their localization properties, several real-valued RBF networks and their supervised learning algorithms have been extended to the Complex domain. In [21, 22], Chen *et al.* first presented the Complex-valued RBF (CRBF) networks, which are direct extensions of the real-valued RBF networks. The structure of such a complex-valued RBF network is presented in Fig. 1.3.

In the figure,

- $\mathbf{z} \in \mathbb{C}^m = [z_1 z_2 \cdots z_m]^T$ are the complex-valued inputs to the network.
- $\mathbf{y}_h \in \mathbb{R}^h = [y_h^1,\ y_h^2,\ \cdots,\ y_h^h]^T$ are the real-valued response of the hidden neurons given by

$$y_h^k = f_g(\mathbf{z}) = exp\frac{(-\mathbf{z}-\mathbf{c})^H(\mathbf{z}-\mathbf{c})}{2\sigma^2};\ k = 1,2,\cdots,h \tag{1.7}$$

where, $\mathbf{c} \in \mathbb{C}^m$ is the m-dimensional complex-valued centers of the Gaussian function in the hidden neurons and $\sigma \in \mathbb{R}$ is the width of the Gaussian function in the hidden neuron and H denotes the Hermitian operator.

[4] For a function $f(z) = u(x,y) + iv(x,y);\ z = x + iy$, the Cauchy Riemann equations are given by:

$$\frac{\partial u}{\partial x} = \frac{\partial v}{\partial y};\ \frac{\partial v}{\partial x} = -\frac{\partial u}{\partial y} \tag{1.6}$$

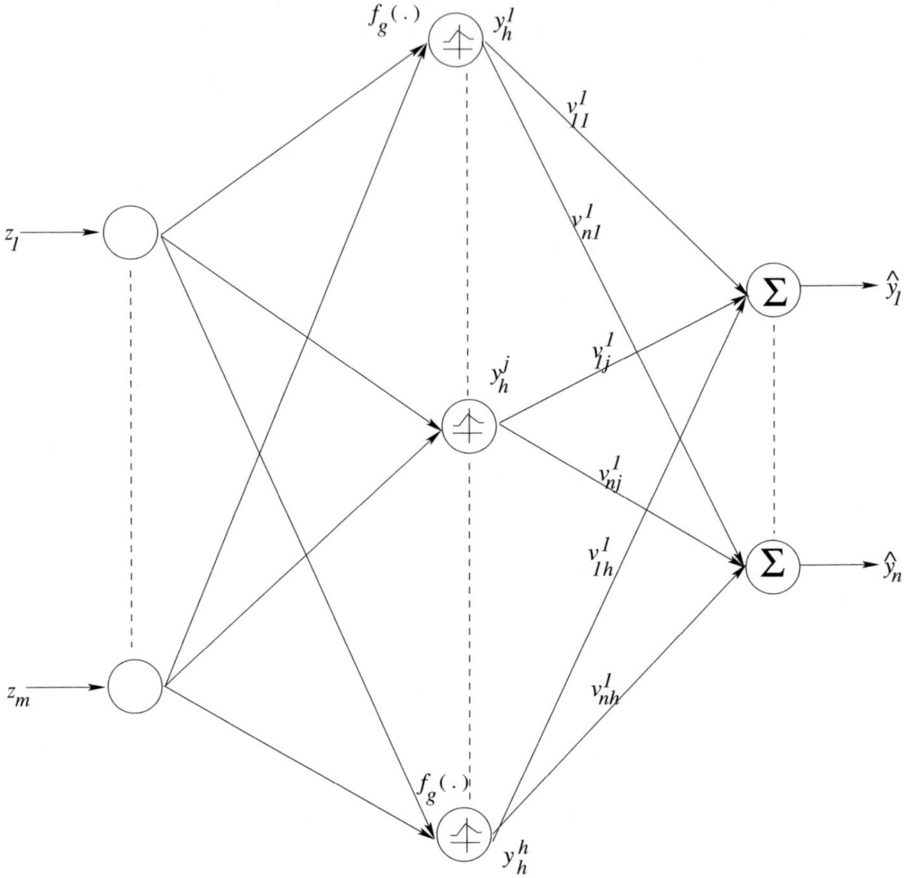

Fig. 1.3 Architecture of a Complex-valued RBF Network

- $V^1 \in \mathbb{C}^{n\times h} = \begin{bmatrix} v_{11}^1 & \cdots & v_{1m}^1 \\ \vdots & & \\ w_{h1}^1 & \cdots & w_{hm}^1 \end{bmatrix}$ are the complex-valued weights connecting the hidden layer and output layer.
- $\widehat{\mathbf{y}} \in \mathbb{C}^n = [\widehat{y}_1, \cdots, \widehat{y}_l, \cdots, \widehat{y}_n]^T$ are the complex-valued outputs of the network, given by: $\widehat{y}_l = \sum_{k=1}^{K} v_{lk}^1 y_h^k$; $l = 1, ..., l, ..., n$.

It can be seen that as the activation function maps $\mathbb{C}^m \to \mathbb{R}$, the responses of the neurons in the hidden layer are real-valued. Similarly, Complex-valued Minimal Resource Allocation Network (CMRAN), presented by Deng *et al*. [8] and Complex-valued Growing and Pruning Network (CGAP-RBF) proposed by Li *et al*. [23] are also direct extensions of the real-valued Minimal Resource Allocation Network (MRAN) proposed by Yingwei *et al*. [24] and Growing and Pruning RBF Network (GAP-RBF) presented by Huang *et al*. [25] respectively. However, in all these algorithms, the Gaussian function that maps $\mathbb{C}^m \to \mathbb{R}$ is used as the basis of the activation function. Hence, despite the centers of the RBF function being complex-valued, the response at the hidden layer is real-valued, resulting in inaccurate phase approximation. Therefore, developing a fully complex-valued RBF network with a fully complex-valued symmetric activation function, capable of better phase approximation, is very important. To address this need, chapter 3 of this book presents a fully complex-valued RBF learning algorithm.

1.2.1.2 Recurrent Neural Networks

A recurrent neural network is a class of neural network where connections between units form a directed cycle. This creates an internal state of the network which allows it to exhibit dynamic temporal behavior. Wang [26] presented a split complex valued recurrent neural network to solve complex-valued linear equations. The complex-valued coefficients were split into their real and imaginary parts and real-valued recurrent neural networks were used to solve the equations. Later, similar to the MLP and RBF framework, the algorithm for the real-valued recurrent neural network [27] is also extended to the Complex domain. Li *et al.* [28] presented a new algorithm for complex-valued recurrent neural network, where each recurrent neuron is modelled as an infinite impulse response filter. Goh and Mandic [29] introduced an augmented complex-valued extended Kalman filter algorithm for the class of nonlinear adaptive filters realized as fully connected recurrent neural networks. The structure of a complex-valued recurrent neural network which consists of N_f neurons, with p external inputs and N feedback connections is shown in Fig. 1.4 [29]. The network has two distinct layers- a feedback layer and a layer of processing elements. Let $\widehat{y}_{l,k}$ denote the complex-valued output of a neuron, $l = 1,...,n$ at time index k and $\mathbf{s}$ be the $(1\times m)$ external complex-valued input vector (i.e., $\mathbf{s} \in \mathbb{C}^m$). The overall input to the network $\mathbf{u}_k$ then represents a concatenation of vectors $\widehat{y}_k$, s and the bias input $(1+i)$, and is given by

$$\mathbf{u}_k = [s_{k-1}, \ \cdots \ s_{k-m}, \ 1+i, \ \widehat{y}_{1,k-1} \ \cdots \ \widehat{y}_{n,k-1}]^T \tag{1.8}$$

For the lth neuron, its weights form a $(m+n+1)\times 1$ dimensional weight vector $\mathbf{v}_l^T = [v_{l,1},\cdots,v_{l,m+n+1}]$, $l = 1,\cdots,N_f$, which are encompassed in the complex-valued weight matrix of the network $V = [\mathbf{v}_1,...,\mathbf{v}_{N_f}]^T$. The output of every neuron can be expressed as

$$y_{l,k} = f_a(net_{l,k}), \ \ l = 1,\cdots,n \tag{1.9}$$

where $f_a(.)$ is a complex-valued nonlinear activation function of a neuron and

$$net_{l,k} = \sum_{d=1}^{m+n+1} w_{ld} u_{d,k} \tag{1.10}$$

is the net input to the lth neuron at time index k, where w_{ld} is the weight connecting the lth output neuron and the dth input neuron and $u_{d,k}$ is the dth input at time index k.

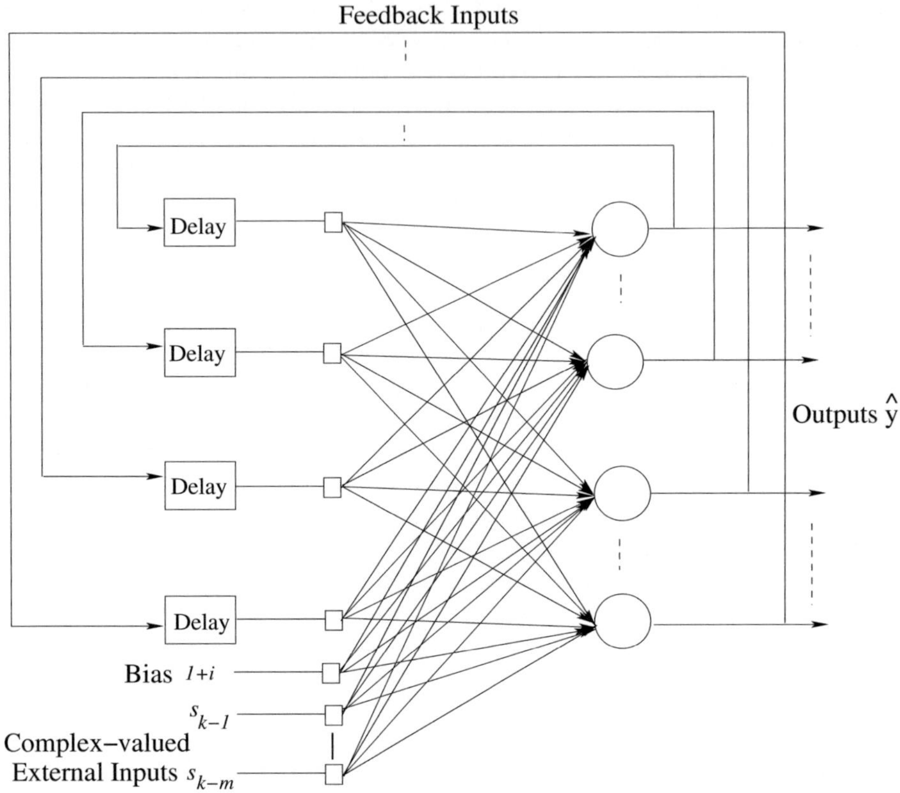

Fig. 1.4 Architecture of a Fully Connected Complex-valued Recurrent Neural Network

Mandic [30] also presented a recurrent neural network based on nonlinear autoregressive moving average models, suitable for processing the generality of complex signals. Later, Zhou and Zurada [31] addressed the boundedness, global attractivity and complete stability of the recurrent neural networks and derived some conditions for those properties. An important application of the complex-valued recurrent neural networks is the estimation of wind profile and wind power as shown by Goh *et al.* [32]. In their work, a complex-valued pipelined recurrent neural network architecture is used, and the network is trained by the complex-valued real-time recurrent learning algorithm with a fully complex-valued activation function to forecast wind signal in its complex form (speed and direction).

1.2.1.3 Error Functions for Supervised Learning

Another area of research interest in CVNN is to identify an efficient error function that minimizes both the magnitude and phase of the complex-valued error signals during learning. The mean squared error function that considers only the magnitude of the complex-valued error is the most commonly used error function. As it is an explicit representation of only the magnitude of the complex-valued error, using this error function results in inaccurate phase approximation. A list of possible choices of error functions for the complex-valued networks is presented by Gangal *et al.* [33]. Chen *et al.* [34] presented a modified error back propagation algorithm for CVNN. They added a term, corresponding to the hidden layer error, to the conventional error function to speed up the learning process. In chapter 2 of this book, we propose a logarithmic error function that uses an explicit representation of both the magnitude and phase of the complex-valued error as an error function for FC-MLP. This makes learning efficient and hence results in more accurate approximation of both the magnitude and phase of the complex-valued signals.

1.2.2 Unsupervised Learning

In unsupervised learning or self-organized learning, there is no external teacher to oversee the learning process [1]. The goal of unsupervised learning techniques is to build representations of the inputs that can be used for decision making, predicting future inputs etc. The Principal Component Analysis (PCA) and the Independent Component Analysis (ICA) are the two commonly used techniques to address this representation. In this section, we discuss the PCA and ICA learning algorithms, available in the literature.

1.2.2.1 Complex-valued Principal Component Analysis

Principal component analysis is an unsupervised learning technique used to reduce the dimensionality of a data set consisting of a large number of interrelated variables, while retaining the variation present in the data as much as possible. In a PCA, the multi-variate data is transformed to a new co-ordinate system using a linear orthogonal transformation.

Rattan and Hsieh [35] extended the real-valued PCA [1] to the Complex-domain. They presented a complex-valued PCA and a Non-linear Complex-valued PCA (NLCPCA) for extraction of nonlinear features from the dataset. The complex-valued neural network model for NLCPCA is an auto-associative feed-forward multi-layer perceptron model. A structure of this NLCPCA is presented in Fig. 1.5 [35].

There are m input and output neurons or nodes corresponding to the m variables. Sandwiched between the input and output layers are 3 hidden layers (starting with the encoding layer, then the bottleneck layer and finally the decoding layer) containing q, 1 and q neurons respectively. The network is composed of two parts: The first part from the input to the bottleneck maps the input z tothe single nonlinear complex principal component $f(\mathbf{z})$. The second part from the bottleneck to the output z_0 is the inverse mapping $g(f(\mathbf{z}))$. Here, the higher dimensional space (m) of the input is reduced linearly to a one-dimensional space at the bottleneck layer given by $f : \mathbb{C}^m \rightarrow \mathbb{C}^1$ and a linear inverse mapping $g : \mathbb{C}^1 \rightarrow \mathbb{C}^m$ maps from bottleneck layer to the m-dimensional output $\mathbf{z}'$, such that the least squares error function (E).

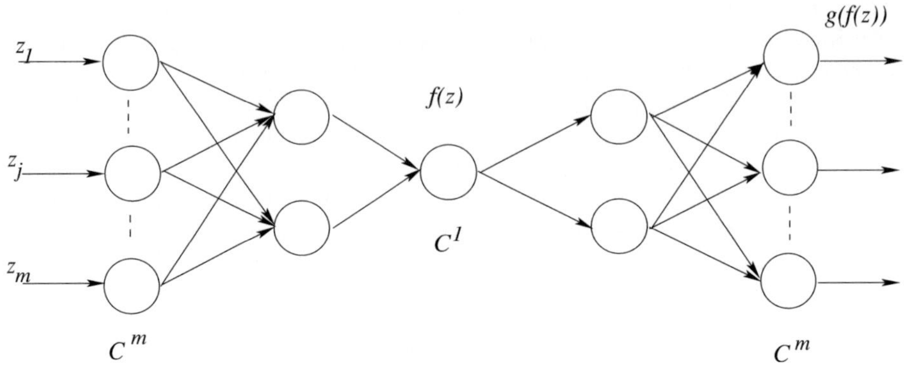

Fig. 1.5 Architecture of a NLCPCA Network

$$E = \sum_{j=1}^{n} \| \mathbf{z}_j - \mathbf{z}'_j \|^2 = \| \mathbf{z}_j - g(f(\mathbf{z}_j)) \|^2 \tag{1.11}$$

is a minimum (with $\mathbf{z}_j$ the jth column of Z). For any input vector $\mathbf{z}$,

$$f(z) = \mathbf{v}^H \mathbf{z} \tag{1.12}$$

where $\mathbf{v}^H$ is the weight vector between the inputs and the bottleneck layer. For auto-associative networks, the target for the output neurons are simply the input data. Increasing the number of neurons in the encoding and decoding layers increases the nonlinear modelling capability of the network.

A review of the various complex-valued principal, minor components/subspace linear/nonlinear rules for complex-weighted neural structures are presented [36]. Several applications of the various unsupervised learning paradigms in the complex domain are also presented in [36].

1.2.2.2 Complex-valued Independent Component Analysis

Independent Component Analysis (ICA) is a signal-processing method used to extract independent sources given only the observed data which is a mixture of the unknown sources. The goal of the ICA can be defined as finding a linear representation of the non-Gaussian data such that the components are statistically independent or as independent as possible [37]. A number of real-valued ICA algorithms are available in the literature and a survey of the real-valued ICA algorithms and their applications are presented in [38].

Complex-valued signals arise frequently in a wide range of applications like communications [39], [40], [41], [42], [43], radar, and biomedicine [44], [45], [46], as most practical modulation formats are of complex type and applications such as radar and magnetic resonance imaging lead to data that are inherently complex-valued. The non-Gaussian nature of the complex-valued data in these applications require the development of the complex-valued ICA algorithms. Bingham *et. al.*, [47, 48] first presented the fast fixed-point type ICA algorithm that is capable of separating the complex-valued linearly mixed source signals, assuming that the original, complex-valued source signals are mutually statistically independent. Fiori *et. al.*, [49] presented an ICA algorithm to solve problems involving complex-valued signals using the maximum-mismatch learning principle. Yang *et. al.*, [50] presented an ICA method for suppression of image and co-channel interference in wireless receivers such that the channel capacity is increased and the receiver's front end is simplified.

The conditions for identifiability, separability and the uniqueness of the linear complex-valued ICA models are established in [51] by extending the well-known condition for the real-valued ICA models. Sallberg *et. al.*, presented the complex-valued ICA with the Kurtosis contrast function in [52]. This ICA algorithm does not exhibit the divergent behavior for Gaussian-only sources that occurs in the Fast ICA method. Later, Li *et. al.*, [53] derived the kurtosis maximization using a gradient update, kurtosis maximization using a fixed-point update, and kurtosis maximization using a Newton update algorithms to perform the complex independent component analysis based on the maximization of the complex kurtosis cost function. The complex maximization of the non-Gaussian cost function (Novey *et. al.*, [54]) and the entropy bound minimization (Li *et. al.*, [55]) are some of the other cost functions used in developing the ICA.

The above works focus on the development and application of the complex-valued ICA to signals of circular sources. The circularity property or properness is an important feature of many complex random signals. At the complex signal level, circularity means that the signal is statistically uncorrelated with its own complex-conjugate. In case of complex random vectors it means that the so-called complementary covariance matrix or pseudo-covariance matrix vanishes. Many widely used signals such as M-QAM and 8-PSK signals and standard complex AWGN possess this circularity property. However, practical imperfections in transmitters and receivers such as I/Q imbalance may cause departures from that property. Moreover, some well known modulation schemes such as BPSK and GMSK are non-circular. Therefore, complex-valued ICA algorithms have been developed to address the non-circularity of the complex-valued signals [56], [57]. As the conventional covariance matrix does not completely describe the second order properties of the non-circular components, Ollilaa *et. al.*, [56] used the generalized uncorrelating transformation instead of the conventional whitening transformation to develop an ICA algorithm for non-circular signals.

The ICA algorithms presented in the above-mentioned papers have been developed for holomorphic functions[5], that satisfy the Cauchy Riemann conditions. The Cauchy Riemann equations were earlier given in Section 1.2.1, Eq. (1.6) and are reproduced here for convenience .

Consider a function

[5] A complex-valued function f(z) of a complex variable z is said to be holomorphic at a point a if it is differentiable at every point within some open disk centered at a.

$$f(z) = u(x,y) + iv(x,y);\ z = x + iy. \tag{1.13}$$

Then the Cauchy Riemann equations are given by:

$$\frac{\partial u}{\partial x} = \frac{\partial v}{\partial y};\ \frac{\partial v}{\partial x} = -\frac{\partial u}{\partial y} \tag{1.14}$$

The holomorphic function $f(z)$ that satisfies the Cauchy Riemann equations requires that the functions $u(x,y)$ and $v(x,y)$ are harmonic[6]. But, some commonly encountered useful functions like

$$f_m(z) = \bar{z} \tag{1.16}$$
$$f_n(z) = \frac{z+\bar{z}}{2} \tag{1.17}$$
$$f_e(z) = |z|^2 = \bar{z}z = x^2 + y^2 \tag{1.18}$$
$$f_p(z) = |z| = \sqrt{\bar{z}z} = \sqrt{x^2+y^2} \tag{1.19}$$

are not harmonic functions and are not differentiable in the standard complex variables sense. It can be noted here that the derivative of the squared error function, which is similar to f_e (Eq. (1.18)), usually employed in gradient descent based/analytical optimization algorithms does not exist in the conventional sense of the complex derivative [58],[59], [60], [61]. However, the functions in Eqs. (1.16), (1.17), (1.18) and (1.19) can be represented in the form of $f(z,\bar{z})$, where they are holomorphic in $z = x+iy$ for fixed $\bar{z}$ and are holomorphic in $z = x-iy$ for fixed z. *i.e.*,

$$\mathbb{R}\text{-derivative of } f(z,\bar{z}) = \frac{\partial f}{\partial z}\,|_{\bar{z}=\text{constant}} \tag{1.20}$$
$$\overline{\mathbb{R}}\text{-derivative of } f(z,\bar{z}) = \frac{\partial f}{\partial \bar{z}}\,|_{z=\text{constant}} \tag{1.21}$$

This fact underlies the development of the $\mathbb{CR}$**- calculus or the Wirtinger calculus** [2], [62]. Eq. (4.12) is called the $\mathbb{R} - derivative$ (the $real - derivative$) and Eq. (4.13) is called the $\overline{\mathbb{R}} - derivative$ (the $conjugate\ \mathbb{R} - derivative$). It is proved in [2] and [11] that the $\mathbb{R} - derivative$ and the $\overline{\mathbb{R}} - derivative$ can be equivalently written as

$$\begin{aligned}\frac{\partial f}{\partial z} &= \frac{1}{2}\left(\frac{\partial f}{\partial x} - j\frac{\partial f}{\partial y}\right)\\ \frac{\partial f}{\partial \bar{z}} &= \frac{1}{2}\left(\frac{\partial f}{\partial x} + j\frac{\partial f}{\partial y}\right)\end{aligned} \tag{1.22}$$

[6] Functions $u(x,y)$ and $v(x,y)$ are harmonic functions, if they satisfy the Laplace equations given by

$$\frac{\partial^2 u(x,y)}{\partial x^2} + \frac{\partial^2 u(x,y)}{\partial y^2} = 0 \text{ and } \frac{\partial^2 v(x,y)}{\partial x^2} + \frac{\partial^2 v(x,y)}{\partial y^2} = 0 \tag{1.15}$$

where the partial derivatives with respect to x and y are *true* partial derivatives of the function $f(z) = f(x,y)$, which is differentiable with respect to x and y. The $\mathbb{R}-derivative$ and the $\overline{\mathbb{R}}-derivative$ are both linear differential operators that obey the product rule of differentiation and the differential rule [2]. Therefore, the $\mathbb{R}-derivative$ of $f_e(z)$ (Eq. (1.18)) with respect to a complex-valued variable a can be derived as:

$$\frac{\partial f_e}{\partial \overline{a}} = \frac{\partial (z\overline{z})}{\partial \overline{a}} = \left(\frac{\partial \overline{z}}{\partial \overline{a}}\right) z + \left(\frac{\partial z}{\partial \overline{a}}\right) \overline{z} \tag{1.23}$$

Thus the pair of partial derivatives of a non-holomorphic functions, defined by Eq. (1.22), is the natural generalization of the single Complex derivative ($\mathbb{C}-derivative$) of a complex-valued holomorphic function. Therefore, the following fact is an easy consequence of the definition in Eq. (1.22), as discussed in [11]:

- A necessary and sufficient condition for a real-valued function $f(z) = f(x,y)$; $z = x+iy$ to have a stationary point with respect to the real-valued parameters $r = (x,y)^T \in \mathbb{R}^2$ is that its $\overline{\mathbb{R}}$ vanishes.
- A necessary and sufficient condition for a real-valued function $f(z) = f(x,y)$; $z = x+iy$ to have a stationary point with respect to the real-valued parameters $\overline{r} = (x,y)^T \in \mathbb{R}^2$ is that its $\mathbb{R}$ vanishes.

This approach that is used to apply the calculus of real variables to make statements about functions of complex variables is known as Wirtinger Calculus. Differentiation of a complex function using the Wirtinger derivatives has been extensively discussed in [63].

Adali *et. al.*, [64] established the theory for Complex Independent Component Analysis (CICA) for nonlinear complex-valued signal processing based on Wirtinger calculus, such that all computations can be directly carried out in the complex domain. Two main approaches for performing ICA within the framework of Wirtinger calculus: maximum likelihood and maximization of non-Gaussianity, have been developed in [64]. The stability of the maximum likelihood complex ICA derived presented in [64] is studied in [64].

1.3 Mode of Learning

In this section, the various complex-valued neural network algorithms available in the literature are classified based on the mode of the learning algorithm. Depending on the sequence in which samples are presented to the network, a learning algorithm can be either a batch-learning algorithm or a sequential learning algorithm.

1.3.1 Complex-valued Batch Learning Algorithms

Batch learning is a learning scheme where all the samples in the training set are presented repeatedly to the network. The free parameters of the network are estimated gradually, over a series of epochs until a specified mean squared error of all the samples is achieved. Such a learning scheme requires the complete training dataset to be available apriori. Several real-valued batch learning algorithms have been extended to the Complex domain. Complex-valued batch learning algorithms are available in the framework of feedforward MLP and RBF networks. Initially, split complex-valued MLP neural networks [3, 65], which treat the complex-valued signals and weights as two real-valued signals were used to operate on complex-valued signals, and the real-valued back propagation algorithm was used to estimate the free parameters of the network. In [66], a neural network based on adaptive activation functions for SC-MLP is presented. But, from the analytical study on sensitivity and initial values of the SC-MLP [5, 67], it is observed that, the process of splitting the complex-valued signals into real and imaginary components introduces phase distortions in complex-valued approximations.

The complex-valued back-propagation algorithm based on the squared error criterion for the complex-valued feedforward network was first derived by Nitta [68], by Benvenuto and Piazza [10], and by Hirose [69]. Ever since, several activation functions have been presented for the feedforward network and the complex-valued back propagation algorithm has been modified accordingly. In [12], Georgiou and Koutsougeras listed the conditions required for a function to serve as a complex-valued activation function and proposed few activation functions for the batch learning MLP network. Later, Kim and Adali [4, 13] relaxed these conditions and suggested elementary transcendental functions as activation functions for FC-MLP. As these functions satisfy Cauchy Riemann conditions, the complex-domain back propagation algorithm was altered with the inclusion of the Cauchy Riemann conditions. Similarly, Chen *et al.* [34] developed a modified error back back-propagation algorithm to solve the problem of local minima, which is inherent to MLP networks. Kim and Guest [70] modified the complex-valued back propagation algorithm to suit to complex-valued signal processing in the frequency domain. You and Hong [14] presented an axially symmetric activation function for solving the QAM equalization problem. However, the axially symmetric activation function is specific to the application considered and choosing the axially symmetric functions for each application is a cumbersome procedure.

There are also other batch learning algorithms in the complex-domain which are direct extensions of their real-valued counterparts. For example, Li *et al.* [16] presented the Complex-valued Extreme Learning Machines (C-ELM), which is the complex-valued extension of the real-valued extreme learning machines. However, all these algorithms compromise either on the analyticity or boundedness or are application specific. Besides, the CBP algorithm presented in all these networks are derived based on the squared error function which does not consider the phase of the complex-valued error explicitly. Hence, one needs to identify an error function which is indicative of both the magnitude and phase error explicitly. In chapter 2 of this book, one such error function is proposed for the complex-valued MLP networks and an Improved Complex-valued MLP (IC-MLP) has been developed in [7] .

On the other hand, although several real-valued RBF learning algorithms [71, 72, 73, 74] have been developed for different applications, only a few of these real-valued neural networks have been extended to the Complex domain. Complex-valued RBF networks, which are direct extensions of the real-valued RBF network, was first presented by Chen *et al.* [21, 22]. For regression problems, [75], Chen *et al.* extended the locally regularized orthogonal least squares of the real-valued RBF network to the Complex domain. Chen [76] presented two training algorithms for symmetrical complex-valued RBF network. These algorithms have been applied to a non-linear beamforming problem. While one of the methods is based on a modified version of the cluster-variation enhanced algorithm, the other method is derived by modifying the orthogonal-forward-selection procedure based on the Fisher ratio of class separability measure. These networks used the Gaussian RBF with Euclidean norm $\|(x-c)\|$ to process complex-valued signals. However, in these networks, the input is not efficiently transmitted to the output, as the activation functions maps $\mathbb{C}^m \rightarrow \mathbb{R}$. Hence, all these networks do not approximate phase accurately. Therefore, it is imperative that a fully complex-valued symmetric function that maps $\mathbb{C}^m \rightarrow C$ be used as a basis for the activation function of a complex-valued RBF network. In chapter 3 of this book, one such function is proposed as a fully complex-valued activation function, based on which a fully complex-valued RBF network is presented and its gradient descent based learning algorithm is derived in [77] .

However, all the aforementioned batch learning algorithms have the following drawbacks:

- *Training dataset:* The batch learning algorithms require the training/testing dataset to be available apriori, hence the network can be trained over several epochs. However, training data may not be available apriori in most real world applications. A few applications like the cancer classification allow temporal changes to the task being learnt.
- *Network structure:* Another critical issue in batch learning algorithms is that the network structure has to be fixed apriori before learning occurs. While fewer neurons in the network result in inaccurate approximation, large network size may result in poor generalization performance and increased computational effort.

These issues in batch learning algorithms motivated the development of sequential learning schemes for neural networks. A few of these sequential learning algorithms are also extended to the Complex domain, and they are discussed briefly in the next section.

1.3.2 Complex-valued Sequential Learning Algorithms

In sequential/online learning algorithms, samples are presented one-by-one and only once to the network. They are discarded after they are presented and learnt by the network. Also, the network structure may evolve during learning, by adding and deleting neurons, as it acquires information from the training sample dataset. Few sequential learning algorithms of the above type have been extended from the Real domain to the Complex domain.

The complex-valued minimal resource allocation network [8] is the first of the kind to be extended from the real-valued minimal resource allocation network [24] sequential learning algorithm. In the CMRAN algorithm, the samples in the training dataset are used to either add a neuron or to delete a neuron, based on the magnitude error of the sample. If neither of these conditions are satisfied, the sample is used to update the parameters of the network using the real-valued extended Kalman filter.

Similarly, the complex-valued growing and pruning (CGAP-RBF) algorithm [23] is another sequential learning algorithm extended from the real-valued growing and pruning RBF [25] learning algorithm. The major differences between the CMRAN and the CGAP-RBF algorithms are:

- In CMRAN, only the sample error and its distance from the nearest neuron is considered for addition of a neuron. However, in the CGAP-RBF, in addition to these parameters, the significance of the sample to the learning accuracy is also considered for addition of a neuron. Thus while the CMRAN uses only the past history of the samples defined by the sliding window, the CGAP-RBF uses the entire past history of the samples.
- While in CMRAN, all the parameters of all the hidden neurons are updated during learning, in the CGAP-RBF algorithm, only the parameters of the nearest neuron are updated.

However, these algorithms are similar in that they use the Gaussian activation function that maps $\mathbb{C}^m \to \mathbb{R}$ and a real-valued EKF for the parameter updates during learning. Though the centers and weights are complex-valued, these algorithms use real-valued Real and Imaginary components of the complex-valued error and weights during the parameter update. Thus, it does not preserve the correlation between the Real/Imaginary components of the complex-valued error/weight information. Recently, Huang *et al.* [78] extended the Incremental Extreme Learning Machine (I-ELM) from the Real domain to the Complex domain. The algorithm randomly adds hidden nodes incrementally and analytically determines the output weights. It has been shown that in spite of the hidden nodes being generated randomly, the network constructed by I-ELM remains as an universal approximator. They show that as long as the hidden layer activation function is complex continuous discriminatory or complex bounded nonlinear piecewise continuous, I-ELM can still approximate any target functions in the complex domain [78]. However, in I-ELM, a few ETFs are used as activation functions at the hidden layer. As the ETFs and their derivatives are known to have their singularities in the finite region of the Complex domain that might interfere with the operating region of the network, identifying a suitable activation function becomes a challenging task. Therefore, there is a need to develop a fully complex-valued sequential learning algorithm, that preserves the complex-valued information, with good generalization performance and approximates phase more accurately. In chapter 7 of this book, we propose one such fully complex-valued sequential learning algorithm called the Complex-valued Self-regulatory Resource Allocation Network [79].

1.4 Applications

In this section, the various applications of the CVNNs are discussed in detail. The applications of CVNNs range from wireless communication to medical imaging. Several complex-valued memories have also been reported in the literature. A complete survey of the various applications of CVNN is discussed by Akira Hirose [80], [81]. In this section, some of the most common applications are highlighted.

1.4.1 Digital Communication: QAM Equalization

Quadrature amplitude modulation is an analog/digital modulation scheme that conveys two message signals by modulating the amplitudes of two carrier waves (quadrature carriers) that are 90° out of phase with each other. Thus, the signals in the QAM schemas are complex-valued. When the QAM signals are transmitted over a channel, the nonlinear characteristics of the channel cause spectral spreading, inter-symbol interference and constellation warping. Hence, an equalizer is essential at the receiver of the communication channel to reduce the precursor inter-symbol interference without any substantial degradation in the signal-to-noise ratio.

In the literature, several works where the complex-valued neural networks are used to solve the QAM equalization problem are reported. Many complex-valued batch learning and sequential learning schemes are used to solve the non-linear, communication channel equalization problems. For example, Chen *et al.* [22] used the batch learning complex-valued RBF network, presented in [21] to solve a digital communication channel equalization. Cha and Kassam [82] used the complex-valued radial basis function network to solve the QAM equalization problem which is considered as a non-linear classification problem. Rajoo Pandey [83] used the complex-valued feed forward neural network for blind equalization with M-ary phase shift keying signals. You and Hong [14] developed an axially symmetric activation function for solving the QAM equalization problem using a complex-valued feed-forward neural network. Deng *et al.* [9], used the sequential learning complex-valued minimal resource allocation network to solve the equalization problem. Li *et al.* [23] used the complex-valued growing and pruning RBF algorithm to solve the equalization of several models, like the Patra model [84], complex-valued linear Chen's model [22], the complex-valued non-linear Cha and Kassam model [82]. As complex-valued neural networks approximate phase more accurately than real-valued neural networks [4, 15, 7], they are more efficient in classifying the complex-valued signals in their quadrants than other real-valued algorithms. In chapter 4 of this book, the different complex-valued neural network based equalizers are used to solve the QAM equalization problem.

1.4.2 Array Signal Processing

The classical problem in array signal processing is to determine the location of an energy radiating planar source relative to the location of the receiver array. Array signal processing comprises of two components viz., Direction of arrival estimation and beam forming. A brief review of the various neural network based antenna array processing is presented by Du *et al.* [85]. As the signals involved in the array signal processing are complex enveloped signals, employing complex-valued neural networks is a wise choice to estimate the direction of arrival and to form the transmitted beam at the receiver.

With the evolution of complex-valued neural networks, they have been employed to solve the array signal processing problems. Bregains and Ares [86] showed that complex-valued neural networks can be incorporated as a very powerful and effective tool in the analysis, synthesis, and diagnostics of antenna arrays. Several works are reported in the literature using complex-valued neural networks for array signal processing applications. For example, Yang *et al.* [87] used the CVNN to solve the array signal processing problem of direction of arrival estimation. Chen *et al.* [75] used a complex-valued RBF network to solve a nonlinear beam forming for multiple-antenna aided communication systems that employ complex-valued quadrature phase shift keying modulation scheme. Chen *et al.* [88] developed a novel complex-valued symmetric radial basis function network based detector, which is capable of approaching the optimal Bayesian performance using channel-impaired training data. They presented a nonlinear beamforming assisted detector for multiple-antenna-aided wireless systems employing complex-valued quadrature phase shift-keying modulation. Similarly, Suksmono and Hirose [89] solved the beamforming problem using the fully complex-valued MLP network. In chapter 4 of this book, we use the fully complex-valued algorithms developed herein to solve the adaptive beam forming problem.

1.4.3 Real-Valued Classification

Nitta [68] showed that the linearly inseparable XOR problem in the Real domain can be easily solved linearly with a single hidden neuron in the Complex domain. As this shows the improved classification ability of the CVNNs, they are also used to solve real-valued classification problems, by phase encoding the real-valued signal, such as the one presented by Amin and Murase [90]. Buchholz and Bihan [91] classified the polarized signals using complex-valued and quaternionic multi-layer perceptron networks. Ozbay *et al.* [92] and Ceylan *et al.* [93] used complex-valued neural networks to classify Doppler signals, which are important in medical applications. On the other hand, Sinha *et al.* [94] used the split complex-valued MLP networks in parallel magnetic resonance image reconstruction which is another important application in medical imaging application.

The multi-layer feed-forward network based on multi-valued neurons has been developed by Aizenberg and Moraga [95]. It is observed that using a traditional architecture of multi-layer feed forward neural network and the high functionality of the multi-valued neurons, it is possible to obtain a new powerful neural network. Its training does not require a derivative of the activation function and its functionality is higher than the functionality of multi-layer feed forward neural network containing the same number of layers and neurons. These advantages of multi-layer feed-forward network based on multi-valued neurons are confirmed by testing using parity n, two spirals, sonar benchmarks and the Mackey-Glass chaotic time series prediction problem. With the introduction of complex-valued multi-valued neurons,

Aizenberg *et al.* [96] extended these multi-valued neurons to solve several real-valued classification problems such as in bio-informatics as shown by Aizenberg and Zurada [95], pattern recognition as shown by [97], blur identification as shown by Aizenberg *et al.* [98] etc.

Amin and Murase [90, 99] presented complex-valued neuron models for real-valued classification problems by phase encoding the real-valued inputs. Activation functions that map complex-valued input to real-valued outputs are presented and the gradient descent based learning rules for the activation functions are derived. It is observed that the classification efficiency of such networks are comparable to that of real-valued networks, and the convergence of these networks are faster than real-valued networks.

1.4.4 Memories

As neural networks are good at learning by example, they are used as memories to learn and recall information/signals [1]. As complex-valued neural networks are also capable of learning and recalling, research focus is also on developing efficient complex-valued logic gates and memory. Complex-valued associative memory, which is a complex-valued Hopfield associative memory, was first introduced by Noest [100]. The capacity of the complex-valued associative memory was improved using the pseudo-relaxation algorithm developed by Kobayashi [101]. Muezzinoglu *et al.* [102] introduced a method to store each element of an integral memory set M subset of 1, 2, ..., $K(n)$ as a fixed point into complex-valued multistate Hopfield network. This method employs a set of inequalities to render each memory pattern as a strict local minimum of a quadratic energy landscape. A novel logic gate is suggested by Kawata and Hirose [103]. This logic gate is capable of learning multiple functions at frequencies different from each other, and analyzing the frequency-domain multiplexing ability in the learning based on complex-valued Hebbian rule. Associative memory based on quaternionic Hopfield network are investigated by Isokawa *et al.* [104]. Quaternion is a class of hypercomplex number systems, and the networks used in [104] are composed of quaternionic neurons and the input, output, threshold and connection weights are all represented in quaternions. The concept of associative memories were then extended to the complex domain by Tanaka Gouhei and Aihara Kazuyuki [105] and are used in gray level image reconstruction.

1.4.5 Other Applications

Another important application of the complex-valued neural networks is the estimation of wind profile and wind power by Goh *et al.* [32]. A complex-valued pipelined recurrent neural network architecture is developed, and the network is trained by the complex-valued real-time recurrent learning algorithm with a fully complex-valued activation function to forecast wind signal in its complex form (speed and direction). The complex-valued neural networks are also used in real-valued image processing applications like image recognition as shown by Pande and Goel [106], blur identification as shown by Aizenberg in [98] etc.

In this chapter, a brief survey of the existing literature on complex-valued neural networks has been presented. The CVNNs can be classified based on various parameters like the nature of signals, the type of learning, the mode of learning and the applications in which they were used. These different classes of complex-valued neural networks have been discussed in this chapter.

References

1. Haykin, S.: Neural Networks: A Comprehensive Foundation. Prentice Hall, New Jersey (1998)
2. Remmert, R.: Theory of Complex Functions. Springer, New York (1991)
3. Leung, H., Haykin, S.: The complex backpropagation algorithm. IEEE Transactions on Signal Processing 39(9), 2101–2104 (1991)
4. Kim, T., Adali, T.: Fully complex multi-layer perceptron network for nonlinear signal processing. Journal of VLSI Signal Processing 32(1/2), 29–43 (2002)
5. Yang, S.-S., Ho, C.-L., Siu, S.: Sensitivity analysis of the split-complex valued multilayer perceptron due to the errors of the i.i.d. inputs and weights. IEEE Transactions on Neural Networks 18(5), 1280–1293 (2007)
6. Zhang, H., Zhang, C., Wu, W.: Convergence of batch split-complex backpropagation algorithm for complex-valued neural networks. Discrete Dynamics in Nature and Society 16, Article ID 329173 (2009), Online Journal, `http://www.hindawi.com/journals/ddns/2009/329173.html`
7. Savitha, R., Suresh, S., Sundararajan, N., Saratchandran, P.: A new learning algorithm with logarithmic performance index for complex-valued neural networks. Neurocomputing 72(16-18), 3771–3781 (2009)
8. Jianping, D., Sundararajan, N., Saratchandran, P.: Complex-valued minimal resource allocation network for nonlinear signal processing. International Journal of Neural Systems 10(2), 95–106 (2000)
9. Deng, J.P., Sundararajan, N., Saratchandran, P.: Communication channel equalization using complex-valued minimal radial basis function neural networks. IEEE Transactions on Neural Networks 13(3), 687–696 (2002)
10. Benvenuto, N., Piazza, F.: On the complex backpropagation algorithm. IEEE Transactions on Signal Processing 40(4), 967–969 (1992)
11. Brandwood, D.H.: A complex gradient operator and its application in adaptive array theory. IEE Proceedings 130, 11–16 (1983)
12. Georgiou, G.M., Koutsougeras, C.: Complex domain backpropagation. IEEE Transactions on Circuits and Systems-II: Analog and Digital Signal Processing 39(5), 330–334 (1992)
13. Kim, T., Adali, T.: Approximation by fully complex multi-layer perceptrons. Neural Computation 15(7), 1641–1666 (2003)
14. You, C., Hong, D.: Nonlinear blind equalization schemes using complex-valued multilayer feedforward neural networks. IEEE Transactions on Neural Networks 9(6), 1442–1455 (1998)
15. Savitha, R., Suresh, S., Sundararajan, N., Saratchandran, P.: Complex-valued function approximation using an improved BP learning algorithm for feed-forward networks. In: IEEE International Joint Conference on Neural Networks (IJCNN 2008), June 1-8, pp. 2251–2258 (2008)

16. Li, M.B., Huang, G.-B., Saratchandran, P., Sundararajan, N.: Fully complex extreme learning machine. Neurocomputing 68(1-4), 306–314 (2005)
17. Huang, G.B., Zhu, Q.Y., Siew, C.K.: Extreme learning machine: a new learning scheme of feedforward neural networks. In: IEEE International Joint Conference on Neural Networks (IJCNN 2004), 25-29, vol. 2, pp. 985–990 (2004)
18. Huang, G.B., Siew, C.K.: Extreme learning machine with randomly assigned RBF kernels. Int. J. Inf. Technol. 11(1) (2005)
19. Kantsila, A., Lehtokangas, M., Saarinen, J.: Complex RPROP-algorithm for neural network equalization of GSM data bursts. Neurocomputing 61, 339–360 (2004)
20. Riedmiller, M., Braun, H.: A direct adaptive method for faster backpropagation learning: The RPROP algorithm. In: Proceedings of the IEEE International Conference on Neural Networks, vol. 1-3, pp. 586–591 (1993)
21. Chen, S., McLaughlin, S., Mulgrew, B.: Complex valued radial basis function network,part I: Network architecture and learning algorithms. EURASIP Signal Processing Journal 35(1), 19–31 (1994)
22. Chen, S., McLaughlin, S., Mulgrew, B.: Complex valued radial basis function network, part II: Application to digital communications channel equalization. Signal Processing 36(2), 175–188 (1994)
23. Li, M.B., Huang, G.B., Saratchandran, P., Sundararajan, N.: Complex-valued growing and pruning RBF neural networks for communication channel equalisation. IEE Proceedings- Vision, Image and Signal Processing 153(4), 411–418 (2006)
24. Yingwei, L., Sundararajan, N., Saratchandran, P.: A sequential learning scheme for function approximation using minimal radial basis function neural networks. Neural Computation 9(2), 461–478 (1997)
25. Huang, G.B., Saratchandran, P., Sundararajan, N.: A generalized growing and pruning RBF (GGAP-RBF) neural network for function approximation. IEEE Transactions on Neural Networks 16(1), 57–67 (2005)
26. Wang, J.: Recurrent neural networks for solving systems of complex-valued linear equations. Electronics Letters 28(18), 1751–1753 (1992)
27. Mandic, D., Chambers, J.: Recurrent Neural Networks for Prediction: Learning Algorithms, Architectures and Stability. John Wiley and Sons, West Sussex (2001)
28. Li, C., Liao, X., Yu, J.: Complex-valued recurrent neural network with IIR neuron model: training and applications. Circuits Systems Signal Processing 21(5), 461–471 (2002)
29. Goh, S.L., Mandic, D.P.: An augmented extended kalman filter algorithm for complex-valued recurrent neural networks. Neural Computation 19(4), 1039–1055 (2007)
30. Mandic, D.P.: Complex valued recurrent neural networks for noncircular complex signals. In: International Joint Conference on Neural Networks (IJCNN 2009), June 14-19, pp. 1987–1992 (2009)
31. Zhou, W., Zurada, J.M.: Discrete-time recurrent neural networks with complex-valued linear threshold neurons. IEEE Transactions on Circuits and Systems 56(8), 669–673 (2009)
32. Mandic, D.P., Javidi, S., Goh, S.L., Kuh, A., Aihara, K.: Complex-valued prediction of wind profile using augmented complex statistics. Renewable Energy 34(1), 196–201 (2009)
33. Gangal, A.S., Kalra, P.K., Chauhan, D.S.: Performance evaluation of complex valued neural networks using various error functions. Proceedings of the World Academy of Science, Engineering and Technology 23, 27–32 (2007)
34. Chen, X.M., Tang, Z., Variappan, C., Li, S.S., Okada, T.: A modified error backpropagation algorithm for complex-valued neural networks. International Journal of Neural Systems 15(6), 435–443 (2005)

35. Rattan, S.S.P., Hsieh, W.W.: Complex-valued neural networks for nonlinear complex principal component analysis. Neural Networks 18(1), 61–69 (2005)
36. Fiori, S.: Nonlinear complex-valued extensions of Hebbian learning: an essay. Neural Computation 17(4), 779–838 (2005)
37. Hyvarinen, A., Karhunen, J., Oja, E.: Independent Component Analysis. John Wiley and Sons, New York (2001)
38. Hyvarinen, A., Oja, E.: Independent component analysis: Algorithms and applications. Neural Networks 13(4-5), 411–430 (2000)
39. Lv, Q., Zhang, X., Jia, Y.: Blind Separation Combined Frequency Invariant Beamforming and ICA for Far-field Broadband Acoustic Signals. In: Wang, J., Liao, X.-F., Yi, Z. (eds.) ISNN 2005. LNCS, vol. 3497, pp. 538–543. Springer, Heidelberg (2005)
40. Chang, A.-C., Jen, C.-W.: Complex-valued ICA utilizing signal-subspace demixing for robust DOA estimation and blind signal separation. Wireless Personal Communications 43(4), 1435–1450 (2007)
41. Lee, I., Kim, T., Lee, T.-W.: Fast fixed-point independent vector analysis algorithms for convolutive blind source separation. Signal Processing 87(8), 1859–1871 (2007)
42. He, Z., Xie, S., Ding, S., Cichocki, A.: Convolutive blind source separation in the frequency domain based on sparse representation. IEEE Transactions on Audio, Speech, and Language Processing 15(5), 1551–1563 (2007)
43. Jen, C.-W., Chen, S.-W., Chang, A.-C.: High-resolution DOA estimation based on independent noise component for correlated signal sources. Neural Computing and Applications 18(4), 381–385 (2008)
44. Calhoun, V.D., Adali, T., Pearlson, G.D., van Zijl, P.C., Pekar, J.J.: Independent component analysis of fMRI data in the complex domain. Magnetic Resonance in Medicine 48(1), 180–192 (2002)
45. Calhoun, V., Adali, T.: Complex infomax: Convergence and approximation of infomax with complex nonlinearities. The Journal of VLSI Signal Processing 44(1-2), 173–190 (2006)
46. Adali, T., Calhoun, V.D.: Complex ICA of brain imaging data. IEEE Signal Processing Magazine 24(5), 136–139 (2007)
47. Bingham, E., Hyvarinen, A.: Ica of complex valued signals: A fast and robust deflationary algorithm. In: International Joint Conference on Neural Networks (IJCNN 2000), vol. 3 (2000)
48. Bingham, E., Hyvarinen, A.: A fast fixed-point algorithm for independent component analysis of complex valued signals. International Journal of Neural Systems 10(1), 1–8 (2000)
49. Fiori, S.: Neural independent component analysis by maximum-mismatch learning principle. Neural Networks 16(8), 1201–1221 (2003)
50. Yang, T., Mikhael, W.B.: A general approach for image and co-channel interference suppression in diversity wireless receivers employing ICA. Circuits, Systems, and Signal Processing 23(4), 317–327 (2004)
51. Erikkson, J., Koivunen, V.: Complex random vectors and ICA models: Identifiability, uniqueness, and separability. IEEE Transactions on Information Theory 52(3), 1017–1029 (2006)
52. Sallberg, B., Grbic, N., Claesson, I.: Complex-valued independent component analysis for online blind speech extraction. IEEE Transactions on Audio, Speech, and Language Processing 16(8), 1624–1632 (2008)
53. Li, H., Adali, T.: A class of complex ICA algorithms based on the kurtosis cost function. IEEE Transactions on Neural Networks 19(3), 408–420 (2008)

54. Novey, M., Adali, T.: Complex ICA by negentropy maximization. IEEE Transactions on Neural Networks 19(4), 596–609 (2008)
55. Li, X.-L., Adali, T.: Complex independent component analysis by entropy bound minimization. IEEE Transactions on Circuits and Systems I 57(7), 1417–1430 (2010)
56. Ollilaa, E., Koivunen, V.: Complex ICA using generalized uncorrelating transform. Signal Processing 89(4), 365–377 (2009)
57. Novey, M., Adali, T.: On extending the complex fast ICA algorithm to noncircular sources. IEEE Transactions on Signal Processing 56(5), 2148–2154 (2008)
58. Brown, J., Churchill, R.: Complex Variables and Applications. McGrawHill, New York (1996)
59. Flanigan, F.: Complex Variables: Harmonic and Analytic Functions. Dover Publications, New York (1983)
60. Le Page, W.: Complex Variables and the Laplace Transforms for Engineers. Dover Publications, New York (1980)
61. Fisher, S.: Complex Variables, 2nd edn. Dover Publications, New York (1999)
62. Wirtinger, W.: Zur formalen theorie der funktionen von mehr komplexen vernderlichen. Annals of Mathematics 97 (1927)
63. Hjorungnes, A., Gesbert, D.: Complex-valued matrix differentiation: Techniques and key results. IEEE Transactions on Signal Processing 55(6), 2740–2746 (2007)
64. Adali, T., Li, H., Novey, M., Cardoso, J.-F.: Complex ICA using nonlinear functions. IEEE Transactions on Signal Processing 56(9), 4536–4544 (2008)
65. Loss, D.V., de Castro, M.C.F., Franco, P.R.G., de Castro, E.C.C.: Phase transmittance RBF neural networks. Electronics Letters 43(16), 882–884 (2007)
66. Uncini, A., Vecci, L., Campolucci, P., Piazza, F.: Complex-valued neural networks with adaptive spline activation function for digital radio links nonlinear equalization. IEEE Transactions on Signal Processing 47(2), 505–514 (1999)
67. Yang, S.S., Siu, S., Ho, C.L.: Analysis of the initial values in split-complex backpropagation algorithm. IEEE Transactions on Neural Networks 19(9), 1564–1573 (2008)
68. Nitta, T.: An extension of the back-propagation algorithm to complex numbers. Neural Networks 10(8), 1391–1415 (1997)
69. Hirose, A.: Continuous complex-valued back-propagation learning. Electronic Letters 28(20), 1854–1855 (1992)
70. Kim, M.S., Guest, C.C.: Modification of back propagation networks for complex-valued signal processing in frequency domain. In: International Joint Conference on Neural Networks (IJCNN 1990), vol. 3, pp. 27–31 (1990)
71. Karim, A., Adeli, H.: Comparison of the fuzzy-wavelet RBFNN freeway incident detection model with the california algorithm. Journal of Transportation Engineering 128(1), 21–30 (2002)
72. Jogensen, T.D., Haynes, B.P., Norlund, C.C.F.: Pruning artificial neural networks using neural complexity measures. International Journal of Neural Systems 18(5), 389–403 (2008)
73. Mayorga, R.V., Carrera, J.: A radial basis function network approach for the computational of inverse continuous time variant functions. International Journal of Neural Systems 17(3), 149–160 (2007)
74. Pedrycz, W., Rai, P., Zurada, J.: Experience-consistent modeling for radial basis function neural networks. International Journal of Neural Systems 18(4), 279–292 (2008)
75. Chen, S., Hong, X., Harris, C.J., Hanzo, L.: Fully complex-valued radial basis function networks: Orthogonal least squares regression and classification. Neurocomputing 71(16-18), 3421–3433 (2008)

76. Chen, S.: Information Science Reference. Complex-valued Neural Networks: Utilizing High-dimensional Parameters, ch. VII. IGI Global snippet, PA (2009)
77. Savitha, R., Suresh, S., Sundararajan, N.: A fully complex-valued radial basis function network and its learning algorithm. International Journal of Neural Systems 19(4), 253–267 (2009)
78. Huang, G.B., Li, M.B., Chen, L., Siew, C.K.: Incremental extreme learning machine with fully complex hidden nodes. Neurocomputing 71(4-6), 576–583 (2008)
79. Suresh, S., Savitha, R., Sundararajan, N.: A sequential learning algorithm for a complex-valued self-regulatory resouce allocation network-csran. IEEE Transactions on Neural Networks 22(7), 1061–1072 (2011)
80. Hirose, A.: Complex-valued neural networks for more fertile electronics. Journal of the Institute of Electronics, Information and Communication Engineers (IEICE) 87(6), 447–449 (2004)
81. Hirose, A.: Complex-valued Neural Networks: Theories and Applications. Series on Innovative Intelligence, vol. 5. World Scientific Publishing Company, Singapore (2004)
82. Cha, I., Kassam, S.A.: Channel equalization using adaptive complex radial basis function networks. IEEE Journal on Selected Areas in Communications 13(1), 122–131 (1995)
83. Pandey, R.: Feedforward neural network for blind equalization with PSK signals. Neural Computing and Applications 14(4), 290–298 (2005)
84. Patra, J.C., Pal, R.N., Baliarsingh, R., Panda, G.: Nonlinear channel equalization for QAM constellation using artificial neural networks. IEEE Transactions on System, Man and Cybernetics, Part B: Cybernetics 29(2), 262–271 (1999)
85. Du, K.L., Lai, A.K.Y., Cheng, K.K.M., Swamy, M.N.S.: Neural methods for antenna array signal processing: A review. Signal Processing 82(4), 547–561 (2002)
86. Bregains, J.C., Ares, F.: Analysis, synthesis and diagnosis of antenna arrays through complex-valued neural networks. Microwave and Optical Technology Letters 48(8), 1512–1515 (2006)
87. Yang, W.H., Chan, K.K., Chang, P.R.: Complex-valued neural network for direction of arrival estimation. Electronics Letters 30(7), 574–575 (1994)
88. Shen, C., Lajos, H., Tan, S.: Symmetric complex-calued RBF receiver for multiple-antenna-aided wireless systems. IEEE Transactions on Neural Networks 19(9), 1659–1665 (2008)
89. Suksmono, A.B., Hirose, A.: Intelligent beamforming by using a complex-valued neural network. Journal of Intelligent and Fuzzy Systems 15(3-4), 139–147 (2004)
90. Amin, M.F., Murase, K.: Single-layered complex-valued neural network for real-valued classification problems. Neurocomputing 72(4-6), 945–955 (2009)
91. Buchholz, S., Bihan, N.L.: Polarized signal classification by complex and quaternionic multi-layer perceptron. International Journal of Neural Systems 18(2), 75–85 (2008)
92. Ozbay, Y., Kara, S., Latifoglu, F., Ceylan, R., Ceylan, M.: Complex-valued wavelet artificial neural network for doppler signals classifying. Artificial Intelligence in Medicine 40(2), 143–156 (2007)
93. Ceylan, M., Ceylan, R., Ozbay, Y., Kara, S.: Application of complex discrete wavelet transform in classification of doppler signals using complex-valued artificial neural network. Artificial Intelligence in Medicine 44(1), 65–76 (2008)
94. Sinha, N., Saranathan, M., Ramakrishna, K.R., Suresh, S.: Parallel magnetic resonance imaging using neural networks. In: IEEE International Conference on Image Processing (ICIP 2007), vol. 3, pp. 149–152 (2007)
95. Aizenberg, I., Moraga, C.: Multilayer feedforward neural network based on multi-valued neurons (MLMVN) and a backpropagation learning algorithm. Soft Computing 11(2), 169–183 (2007)

96. Aizenberg, I., Moraga, C., Paliy, D.: A feedforward neural network based on multi-valued neurons. In: Computational Intelligence, Theory and Applications. Advances in Soft Computing, XIV, pp. 599–612. Springer, Berlin (2005)
97. Aizenberg, I., Aizenberg, N.: Pattern Recognition Using Neural Network Based on Multi-Valued Neurons. In: Mira, J. (ed.) IWANN 1999. LNCS, vol. 1607, pp. 383–392. Springer, Heidelberg (1999)
98. Aizenberg, I., Paliy, D.V., Zurada, J.M., Astola, J.T.: Blur identification by multi-layer neural network based on multivalued neurons. IEEE Transactions on Neural Networks 19(5), 883–898 (2008)
99. Amin, M.F., Islam, M.M., Murase, K.: Ensemble of single-layered complex-valued neural networks for classification tasks. Neurocomputing 72(10-12), 2227–2234 (2009)
100. Noest, A.J.: Phasor neural networks. Neural Information Processing Systems 2, 584–591 (1989), Online Journal, `http://books.nips.cc/nips02.html`
101. Kobayashi, M.: Pseudo-relaxation learning algorithm for complex-valued associative memory. International Journal of Neural Systems 18(2), 147–156 (2008)
102. Muezzinoglu, M.K., Guzelis, C., Zurada, J.M.: A new design method for the complex-valued multistate Hopfield associative memory. IEEE Transactions on Neural Networks 14(4), 891–899 (2003)
103. Kawata, S., Hirose, A.: Frequency-multiplexing ability of complex-valued Hebbian learning in logic gates. International Journal of Neural Systems 18(2), 173–184 (2008)
104. Isokawa, T., Nishimura, H., Kamiura, N., Matsui, N.: Associative memory in quaternionic Hopfield neural network. International Journal of Neural Systems 18(2), 135–145 (2008)
105. Tanaka, G., Aihara, K.: Complex-valued multistate associative memory with nonlinear multilevel functions for gray-level image reconstruction. IEEE Transactions on Neural Networks 20(9), 1463–1473 (2009)
106. Pande, A., Goel, V.: Complex-valued neural network in image recognition: A study on the effectiveness of radial basis function. Proceedings of World Academy of Science, Engineering and Technology 20, 220–225 (2007)

Chapter 2
Fully Complex-valued Multi Layer Perceptron Networks

This chapter focuses specifically on the study of different Complex-valued MLP (CMLP) networks and their learning algorithms in detail. Complex-valued MLPs can be broadly classified into two types depending on the way in which the complex-valued signal is handled. They are viz., Split Complex-valued MLP (SC-MLP) and Fully Complex-valued MLP (FC-MLP). In a split complex-valued MLP, the complex-valued inputs and outputs are split into two real-valued inputs using rectangular or polar coordinate systems, though the rectangular co-ordinate based splitting is the most commonly used one. On the other hand, FC-MLP neural networks process the complex-valued input signals, using fully complex-valued activation functions and weights to give fully complex-valued output signals.

First, we present the SC-MLP network, where the complex-valued inputs and outputs are split into two real-valued inputs/outputs, i.e., real and imaginary values of complex-valued inputs/outputs. Here, the well-known real-valued MLP and its real-valued gradient descent based learning algorithm are used to update the free parameters of the network. After learning is complete, the complex-valued output ($\widehat{\mathbf{y}}$) is reconstructed by a complex summation of the two output nodes ($\widehat{\mathbf{y}}_R + i\widehat{\mathbf{y}}_I$). However, SC-MLP uses real-valued representation of the complex-valued signals and weights, the gradients used in the learning algorithm are also real-valued which do not reflect the true complex-valued gradient.

Next, we present an FC-MLP network that uses complex-valued activation functions and complex-valued weights. As all the signals and weights of an FC-MLP are complex-valued, the gradient of such a network represents the true gradient of the complex-valued function and hence improves the phase approximation ability. But, the challenge in an FC-MLP is the proper selection of an activation function, which is analytic and bounded almost everywhere. This is because, according to Liouville's theorem, a bounded entire function must be a constant in $\mathbb{C}$ [1]. In [1], a set of elementary transcendental functions like "*asinh*", "*atan*", "*cos*", "*sin*", "*atanh*", "*asin*", "*tan*", "*tanh*" etc, which are *almost everywhere* bounded functions were suggested as possible choices of activation functions for FC-MLP and the learning algorithm was also derived. It was observed from the complex-valued gradient based update rule that the complex-valued gradients of the free parameters are similar to that of

S. Suresh et al.: Supervised Learning with Complex-valued Neural Networks, SCI 421, pp. 31–47.
springerlink.com

the real-valued gradient based updates, except for the complex conjugate of the activation functions and the inputs of each layer. Although there are other activation functions reported in the literature for FC-MLP, they are mostly specific to the applications considered [2] and only ETFs are studied in this book. As the singularities of the ETFs and their derivatives strongly influence the performance of a FC-MLP, it is necessary to identify a new activation function which is analytic and bounded almost everywhere and one that does not interfere with the region of operation of the network. Moreover, FC-MLP uses the mean-squared error function that represents only the magnitude of the complex-valued error explicitly. Hence, there is a need to identify an error function that explicitly represents both the magnitude and phase, one which approximates both magnitude and phase explicitly in a simultaneous way. These issues are addressed in this chapter with a new proposal of an Improved Complex-valued MLP (IC-MLP) with an '*exp*' activation function and a logarithmic error function.

2.1 Complex-valued Multi-Layer Perceptron Networks

In this section, a description of existing complex-valued MLP networks in the literature are presented along with their problems.

2.1.1 Split Complex-valued Multi-Layer Perceptron

Let $\{(\mathbf{z}_1, \mathbf{y}_1), \cdots (\mathbf{z}_t, \mathbf{y}_t), \cdots (\mathbf{z}_N, \mathbf{y}_N)\}$ be the observation data/training data, where $\mathbf{z}_t \in \mathbb{C}^m$ is the m-dimensional input features of observation t,$\mathbf{y}_t \in \mathbb{C}^n$ is the n-dimensional target of observation t. For notational convenience, the subscript t is dropped in future discussions.

Let $\mathbf{z} = [z_1 \; z_2 \cdots z_m]^T$ be the complex-valued signal, split into its real and imaginary components as: $\mathbf{z} = \mathbf{z}_R + i\mathbf{z}_I$, where $\mathbf{z}_R \in \mathbb{R}^m$ is the real part of the m-dimensional input feature, and $\mathbf{z}_I \in \mathbb{R}^m$ is the imaginary part of the m-dimensional input feature. Similarly, the target vector can be split into $\mathbf{y} = \mathbf{y}_R + i\mathbf{y}_I$, where $\mathbf{y}_R \in \mathbb{R}^n$ is the real part of the n dimensional target vector, and $\mathbf{y}_I \in \mathbb{R}^n$ is the imaginary part of the n dimensional target vector. The m-dimensional complex-valued input vector $\mathbf{z}$ is split into its real and imaginary parts and the $2m$ dimensional input signal is used as input to SC-MLP. Thus, the input to SC-MLP network is described by: $\mathbf{x} \in \mathbb{R}^{2m} = [\mathbf{z}_R \quad \mathbf{z}_I]^T$ and the targets to the network described by: $\mathbf{o} \in \mathbb{R}^{2n} = [\mathbf{y}_R \quad \mathbf{y}_I]^T$ and the network output described by: $\widehat{\mathbf{o}} \in \mathbb{R}^{2n} = [\widehat{\mathbf{y}}_R \quad \widehat{\mathbf{y}}_I]^T$.

Thus, the split complex-valued MLP network uses the same architecture as that of the real-valued MLP. In general, a multi layer perceptron has an input layer, hidden layers and an output layer. In this book, we consider only a single hidden layer network. The architecture of the single hidden layer split complex-valued MLP network is shown in Fig. 2.1. From the figure, we can see that the architecture has $2m$ input neurons in the input layer, h hidden neurons in the hidden layer, and $2n$ output neurons in the output layer. Hence, the network can be represented by $\mathcal{N}^{2m:h:2n}$. In SC-MLP discussed in this chapter, the hidden neurons use *bipolar sigmoidal*

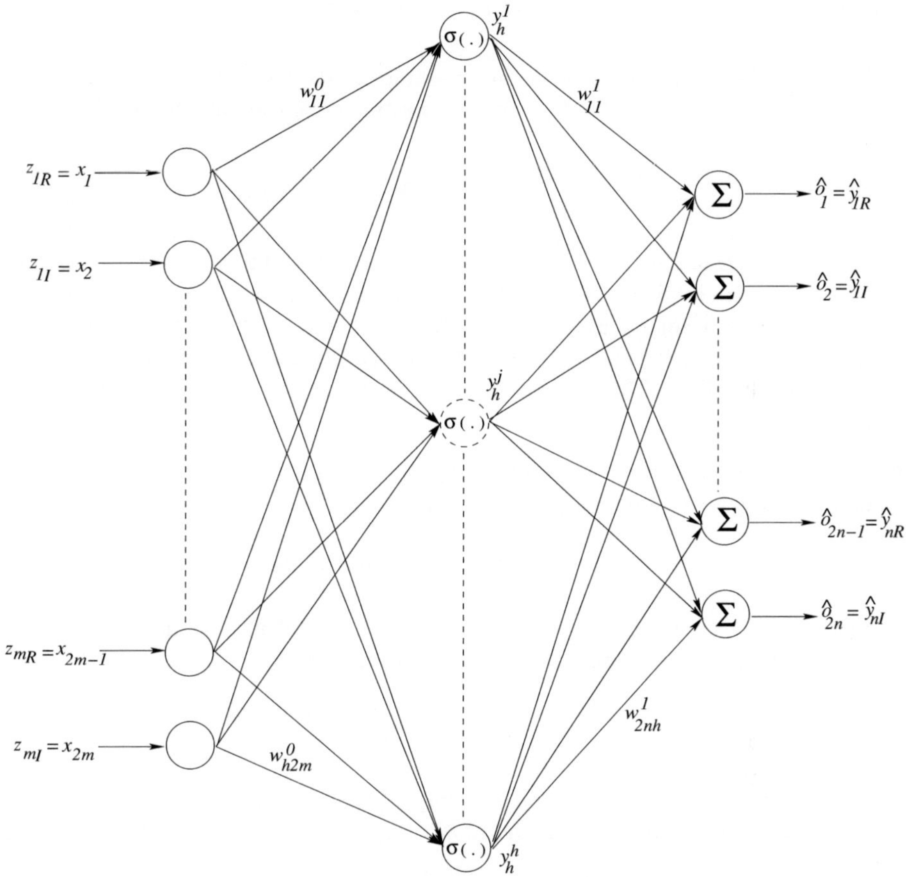

Fig. 2.1 Architecture of an SC-MLP Network

activation and the output neuron use a linear activation function. The output of the kth hidden neuron y_h^k is given by

$$y_h^k = \sigma\left(\sum_{j=1}^{2m} w_{kj}^0 x_j\right); \quad k = 1,\ 2,\ \cdots\ h \tag{2.1}$$

where w_{kj}^0 is the real-valued input weight connecting the kth hidden neuron and the jth input neuron, and $\sigma(.)$ is the real-valued *bipolar sigmoidal* activation function $(\sigma(.) : \mathbb{R}^m \to \mathbb{R})$.

Similarly, the predicted output of the kth output neuron of SC-MLP $\widehat{o}_k$ is given by

$$\widehat{o}_k = \sum_{j=1}^{h} w_{kj}^1 y_h^j; \quad k = 1,\ 2,\ \cdots\ 2n \tag{2.2}$$

where w^1_{kj} is the real-valued output weight connecting the jth hidden neuron and the kth output neuron.

The objective is to estimate the free parameters (W^0 and W^1) of the network, such that the error is minimized. To ensure this, the error function defined by

$$E = \frac{1}{2}\sum_{k=1}^{n} e_k^2 \text{ where } e_k = o_k - \widehat{o}_k \tag{2.3}$$

is used. Minimizing the error function is achieved by updating the input and output weights for each sample of the training data set repeatedly until the error converges to the required training accuracy or over fixed number of epochs. The adjustments of weight matrices (W^0 and W^1) are done in accordance with the standard back propagation algorithm. The gradient based error correction rule for the weights are given below:

- The gradient based error correction Δw^1_{ij} for the weight connecting the jth hidden neuron and the kth output neuron is given by

$$\Delta w^1_{kj} = \eta \delta^1_k y^j_h; \qquad \text{where} \quad \delta^1_k = e_k \tag{2.4}$$

- Similarly, for the weight w^0_{kj} connecting the kth hidden neuron and the jth input neuron, the gradient based error correction is given by

$$\Delta w^0_{kj} = \eta \delta^0_k x_j; \qquad \text{where} \quad \delta^0_k = \sigma'(W^0\mathbf{x}) \sum_l w^{1'}_{lk} e_{l'} \tag{2.5}$$

Since back propagation is a batch learning algorithm, the training samples are presented for a number of times (number of epochs) till the desired accuracy is achieved.

Finally, the actual complex-valued output is reconstructed as the Complex sum of the Real and Imaginary components of the predicted output.

$$\widehat{y}_k = \widehat{o}_k + i\widehat{o}_{n+k}; \; k = 1, 2, \cdots n \tag{2.6}$$

Another common method is to split the complex-valued inputs and outputs based on polar co-ordinates, that is, the magnitude and phase of the complex-valued signals. The real-valued MLP is used in this method, and the learning algorithm also follows the real-valued back propagation algorithm.

Issues in SC-MLP: Either way, the structure of the network is: $\mathscr{N}^{2m:h:2n}$. Usually, $h > (2m+2n)$. Thus, it is computationally expensive to estimate the parameters of an SC-MLP network. Besides, in an SC-MLP network, the gradients are real-valued and hence, do not reflect the true complex-valued gradients. Also, the convergence of an SC-MLP network is highly sensitive to the choice of the learning rate parameter.

In the next section, a FC-MLP [1] network is presented in detail.

2.1.2 Fully Complex-valued Multi-Layer Perceptron

Fully complex-valued MLP (FC-MLP) networks operate on the complex-valued signal as a single entity. The architecture of an FC-MLP is similar to its real-valued counterpart except that the inputs/outputs, weights and activation functions are complex-valued. As described earlier, let $\{(\mathbf{z}_1, \mathbf{y}_1), \cdots (\mathbf{z}_t, \mathbf{y}_t), \cdots (\mathbf{z}_N, \mathbf{y}_N)\}$ be the set of complex-valued inputs and outputs of the training data set, where $\mathbf{z}_t \in \mathbb{C}^m$ is the m dimensional complex-valued inputs and $\mathbf{y}_t \in \mathbb{C}^n$ is the n dimensional complex-valued targets to FC-MLP network.

FC-MLP operates on the m dimensional complex-valued inputs (subscript t is dropped for notational convenience) $\mathbf{z} = [z_1 \ z_2 \cdots z_m]$, using the complex-valued activation function $f_a(.)$, complex-valued input weight matrix V^0 and complex-valued output weight matrix V^1 to generate the n dimensional complex-valued outputs $\widehat{\mathbf{y}} = [\widehat{y}_1 \ \widehat{y}_2 \cdots y_n]$. The architecture of a three layered FC-MLP network is shown in Fig. 2.2.

FC-MLP network has m input neurons, h hidden neurons and n output neurons and is represented by $\mathcal{N}^{m:h:n}$. The activation function, at the hidden and output layer of the FC-MLP neural network is also fully complex-valued. The selection of an appropriate complex-valued activation function is a very challenging problem in the context of a FC-MLP neural network because according to Liouville's theorem, an entire and bounded function is a constant function in the Complex domain. Hence, the conditions for a fully complex-valued activation have been relaxed and reduced in [1] as: *In a bounded domain of a complex plane $\mathbb{C}$, a fully complex nonlinear activation function f(z) needs to be analytic and bounded almost everwhere.* In [1], a set of elementary transcendental functions were also suggested as fully complex-valued activation function for a fully complex-valued MLP. The list of possible ETF function are given in chapter 2, section 1.1.2.1. Let $f_a : \mathbb{C}^m \to \mathbb{C}$ be the activation function,(that is, one of the ETF's presented in 1.1.2.1). Then,

$$f_a(z) = f_a(z_R + iz_I) \tag{2.7}$$

where $z = z_R + iz_I$. Then, the output of the jth hidden neuron of an FC-MLP network is given by

$$z_h^k = f_a\left(\sum_{j=1}^{m} v_{kj}^0 z_j\right); \ k = 1, 2, \cdots h \tag{2.8}$$

where v_{kj}^0 is the complex-valued weight connecting the kth input neuron and the jth hidden neuron.

Similarly, the output of the lth output neuron of the network is given by

$$\widehat{y}_l = f_a\left(\sum_{k=1}^{h} v_{lk}^1 z_h^k\right); \ l = 1, 2, \cdots n \tag{2.9}$$

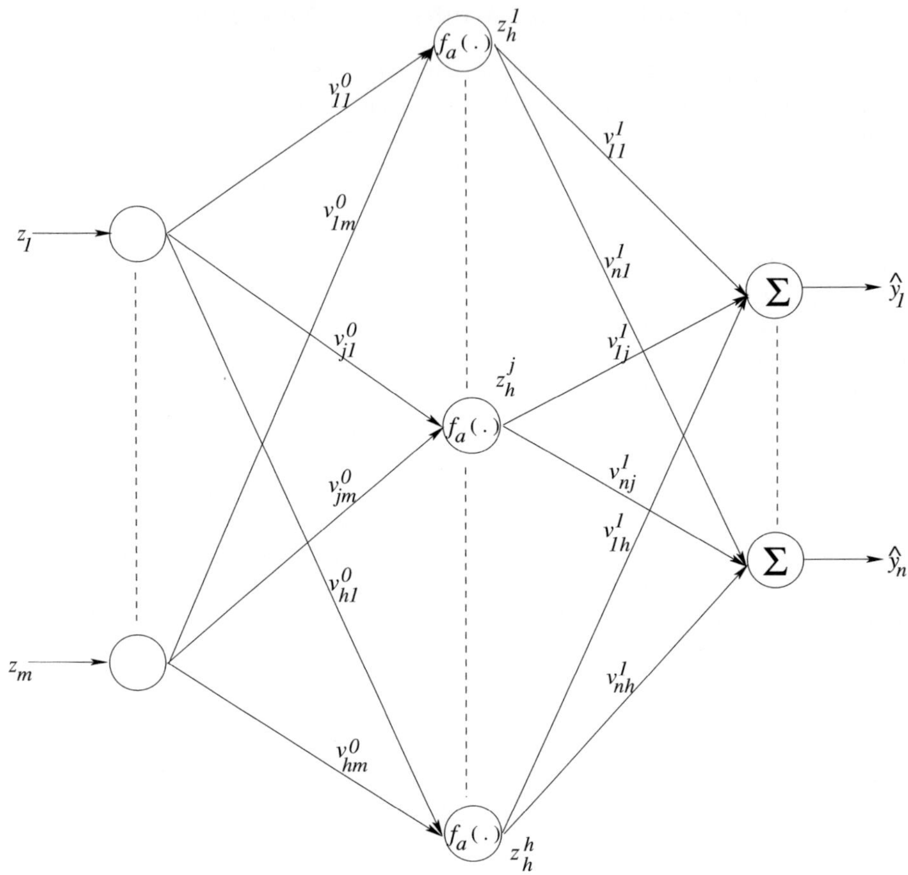

Fig. 2.2 Architecture of an FC-MLP Network

where v_{lk}^1 is the complex-valued weight connecting the kth hidden neuron and the lth output neuron.

The objective, here is to estimate the free parameters of the network, V^0 and V^1 such that the error is minimized

$$E = \frac{1}{2}\mathbf{e}^H\mathbf{e} = \frac{1}{2}\sum_{k=1}^{n} e_k\overline{e}_k; \tag{2.10}$$

where $e_k = y_k - \widehat{y}_k$, $\overline{e}_k$ is the complex conjugate of e_k and H denotes the complex Hermitian operator. The free parameters of FC-MLP network are estimated using the complex-valued back propagation algorithm, which was discussed in detail in [1] and has been reproduced here for completeness.

Assuming that the complex-valued function considered satisfies the relaxed desired conditions needed for a fully complex-valued activation function, that is, if it is analytic *almost everywhere*, the Cauchy Riemann equations defined by

$$\begin{aligned}
&\text{If } f_a(z) = p(z_R,z_I) + i\,q(z_R,z_I), \text{ where } z = z_R + i\,z_I,\\
&p(z_R,z_I) \text{ and } q(z_R,z_I) \text{are the real and imaginary parts}\\
&\qquad \text{of the complex-valued activation function} f_a(z)\\
&\text{then, } \frac{\partial p}{\partial z_R} = \frac{\partial q}{\partial z_I} \text{ and } \frac{\partial q}{\partial z_R} = -\frac{\partial p}{\partial z_I}
\end{aligned} \tag{2.11}$$

can be used to simplify the fully complex-valued back propagation algorithm, originally derived in [3]. The simplification uses the following expression:

$$f_a'(z) = f_{z_R} = -if_{z_I} \tag{2.12}$$

where f_{z_R} and f_{z_I} are partial derivatives with respect to Real and Imaginary components, z_R and z_I, respectively. Here, If $z = z_R + iz_I$ and $f_a(z) = p(z_R,z_I) + iq(z_R,z_I)$,

$$\frac{\partial f_a}{\partial z} = \frac{1}{2}\left(\frac{\partial f_a}{\partial z_R} - i\frac{\partial f_a}{\partial z_I}\right). \tag{2.13}$$

With these definitions, the complex-valued back propagation learning algorithm can be derived as given below.

The output of the l-th output neuron of FC-MLP can be written as:

$$\begin{aligned}
\widehat{y}^l &= \sum_k v^1_{lk} z^k_h;\ l = 1,\cdots,n\\
\text{Let } z^k_h &= p_k + i\,q_k\\
\text{Then, } \widehat{y}^l &= \sum_k (v^{1^R}_{lk} + i\,v^{1^I}_{lk})\,(p_k + i\,q_k)\\
\Longrightarrow \widehat{y}^R_l + i\widehat{y}^I_l &= \sum_k (v^{1^R}_{lk}\,p_k - v^{1^I}_{lk} q_k) + i(v^{1^R}_{lk}\,q_k + v^{1^I}_{lk} p_k)
\end{aligned} \tag{2.14}$$

where the superscripts R and I represent the real and imaginary components of the complex-valued signals, respectively.

The derivative of the error function with respect to the output weight v^1_{lk} is given by:

$$\frac{\partial E}{\partial v^1_{lk}} = \nabla v^1_{lk}\,E = \frac{\partial E}{\partial v^{1^R}_{lk}} + i\frac{\partial E}{\partial v^{1^I}_{lk}} \tag{2.15}$$

$$\frac{\partial E}{\partial v^{1^R}_{lk}} = \frac{\partial E}{\partial \widehat{y}^R_l}\frac{\partial \widehat{y}^R_l}{\partial v^{1^R}_{lk}} + \frac{\partial E}{\partial \widehat{y}^I_l}\frac{\partial \widehat{y}^I_l}{\partial v^{1^R}_{lk}} \tag{2.16}$$

$$\frac{\partial E}{\partial v^{1^I}_{lk}} = \frac{\partial E}{\partial \widehat{y}^R_l}\frac{\partial \widehat{y}^R_l}{\partial v^{1^I}_{lk}} + \frac{\partial E}{\partial \widehat{y}^I_l}\frac{\partial \widehat{y}^I_l}{\partial v^{1^I}_{lk}} \tag{2.17}$$

Letting $\frac{\partial E}{\partial \widehat{y}^R_l} = \delta^R_l$ and $\frac{\partial E}{\partial \widehat{y}^I_l} = \delta^I_l$, we have,

$$\frac{\partial E}{\partial v_{lk}^{1^R}} = \delta_l^R \frac{\partial \widehat{y}_l^R}{\partial v_{lk}^{1^R}} + \delta_l^I \frac{\partial \widehat{y}_l^I}{\partial v_{lk}^{1^R}} \tag{2.18}$$

$$\frac{\partial E}{\partial v_{lk}^{1^I}} = \delta_l^R \frac{\partial \widehat{y}_l^R}{\partial v_{lk}^{1^I}} + \delta_l^I \frac{\partial \widehat{y}_l^I}{\partial v_{lk}^{1^I}} \tag{2.19}$$

From Eq. (2.14), we can see that

$$\frac{\partial \widehat{y}_l^R}{\partial v_{lk}^{1^R}} = p_k; \ \frac{\partial \widehat{y}_l^R}{\partial v_{lk}^{1^I}} = -q_k; \ \frac{\partial \widehat{y}_l^I}{\partial v_{lk}^{1^R}} = q_k; \ \frac{\partial \widehat{y}_l^I}{\partial v_{lk}^{1^I}} = p_k \tag{2.20}$$

Substituting the inferences from Eq. (2.20) into Eqs. (2.18) and (2.22),

$$\frac{\partial E}{\partial v_{lk}^{1^R}} = \delta_l^R \ (p_k) + \delta_l^I \ (q_k) \tag{2.21}$$

$$\frac{\partial E}{\partial v_{lk}^{1^I}} = \delta_l^R \ (-q_k) + \delta_l^I \ (p_k) \tag{2.22}$$

Hence,

$$\begin{aligned}\frac{\partial E}{\partial v_{lk}^{1}} &= \frac{\partial E}{\partial v_{lk}^{1^R}} + i \frac{\partial E}{\partial v_{lk}^{1^I}} = (\delta_l^R \ (p_k) + \delta_l^I \ (q_k)) + i(\delta_l^R \ (-q_k) + \delta_l^I \ (p_k)) \\ &= \delta_l^R (p_k - \ i \ q_k) + \delta_l^I (i \ p_k + q_k) \ = \ \delta_l^R (p_k - i \ q_k) + i \delta_l^I (i \ p_k - i \ q_k) \\ &= (\delta_l^R + \ i \ \delta_l^I) \ (p_k - i \ q_k) \implies \frac{\partial E}{\partial v_{lk}^{1}} \ = \ - \ \delta_l \ \overline{z}_h^k \end{aligned} \tag{2.23}$$

Therefore, the update rule for the output weight connecting the k-th hidden neuron and the l-th output neuron is given by:

$$\Delta v_{lk} = \eta_v \ \delta_l \ \overline{z}_h^k \tag{2.24}$$

where $\delta_l \ = \ y_l \ - \ \widehat{y}_l$ and η_v is the learning rate that can be a real-valued or complex-valued scalar.

Similarly, the input weight update rule can be derived from the first derivative of the mean squared error function in Eq. (2.10) with respect to the input weights v_{kj}^0; $k = 1, \cdots, h$, $j = 1, cdots, m$. For this purpose, let us consider

$$z_h^k = a_k + i \ b_k \ = \ f_a \left(\sum_{j=1}^{m} v_{kj}^0 \ z_j \right) \ = \ f_a \left(p_k + \ i \ q_k\right)$$

$$\text{where} p_k + \ i \ q_k = \sum_{j=1}^{m} v_{kj}^0 \ z_j \ = \ \sum_{j=1}^{m} \left(v_{kj}^{0^R} + i v_{kj}^{0^I}\right) \left(z_j^R \ + \ i \ z_j^I\right)$$

$$\implies p_k + \ i \ q_k = \sum_{j=1}^{m} \left(v_{kj}^{0^R} . z_j^R \ - \ v_{kj}^{0^I} . z_j^I\right) + i \left(v_{kj}^{0^I} . z_j^R \ + \ v_{kj}^{0^R} . z_j^I\right) \tag{2.25}$$

where the superscripts R and I represent the real and imaginary components of the complex-valued signals, respectively.

The first derivative of the mean squared error function with respect to the input weights v_{kj}^0 is given by:

$$\frac{\partial E}{\partial v_{kj}^0} = \frac{\partial E}{\partial v_{kj}^{0^R}} + i\frac{\partial E}{\partial v_{kj}^{0^I}} \tag{2.26}$$

$$\begin{aligned}\frac{\partial E}{\partial v_{kj}^{0^R}} = \frac{\partial E}{\partial \widehat{y}_l^R}\left[\frac{\partial \widehat{y}_l^R}{\partial a_k}\frac{\partial a_k}{\partial p_k}\frac{\partial p_k}{\partial v_{kj}^{0^R}} + \frac{\partial \widehat{y}_l^R}{\partial a_k}\frac{\partial a_k}{\partial q_k}\frac{\partial q_k}{\partial v_{kj}^{0^R}} + \frac{\partial \widehat{y}_l^R}{\partial b_k}\frac{\partial b_k}{\partial p_k}\frac{\partial p_k}{\partial v_{kj}^{0^R}} + \frac{\partial \widehat{y}_l^R}{\partial b_k}\frac{\partial b_k}{\partial q_k}\frac{\partial q_k}{\partial v_{kj}^{0^R}}\right] \\ + \frac{\partial E}{\partial \widehat{y}_l^I}\left[\frac{\partial \widehat{y}_l^I}{\partial a_k}\frac{\partial a_k}{\partial p_k}\frac{\partial p_k}{\partial v_{kj}^{0^R}} + \frac{\partial \widehat{y}_l^I}{\partial a_k}\frac{\partial a_k}{\partial q_k}\frac{\partial q_k}{\partial v_{kj}^{0^R}} + \frac{\partial \widehat{y}_l^I}{\partial b_k}\frac{\partial b_k}{\partial p_k}\frac{\partial p_k}{\partial v_{kj}^{0^R}} + \frac{\partial \widehat{y}_l^I}{\partial b_k}\frac{\partial b_k}{\partial q_k}\frac{\partial q_k}{\partial v_{kj}^{0^R}}\right]\end{aligned} \tag{2.27}$$

$$\begin{aligned}\frac{\partial E}{\partial v_{kj}^{0^I}} = \frac{\partial E}{\partial \widehat{y}_l^R}\left[\frac{\partial \widehat{y}_l^R}{\partial a_k}\frac{\partial a_k}{\partial p_k}\frac{\partial p_k}{\partial v_{kj}^{0^I}} + \frac{\partial \widehat{y}_l^R}{\partial a_k}\frac{\partial a_k}{\partial q_k}\frac{\partial q_k}{\partial v_{kj}^{0^I}} + \frac{\partial \widehat{y}_l^R}{\partial b_k}\frac{\partial b_k}{\partial p_k}\frac{\partial p_k}{\partial v_{kj}^{0^I}} + \frac{\partial \widehat{y}_l^R}{\partial b_k}\frac{\partial b_k}{\partial q_k}\frac{\partial q_k}{\partial v_{kj}^{0^I}}\right] \\ + \frac{\partial E}{\partial \widehat{y}_l^I}\left[\frac{\partial \widehat{y}_l^I}{\partial a_k}\frac{\partial a_k}{\partial p_k}\frac{\partial p_k}{\partial v_{kj}^{0^I}} + \frac{\partial \widehat{y}_l^I}{\partial a_k}\frac{\partial a_k}{\partial q_k}\frac{\partial q_k}{\partial v_{kj}^{0^I}} + \frac{\partial \widehat{y}_l^I}{\partial b_k}\frac{\partial b_k}{\partial p_k}\frac{\partial p_k}{\partial v_{kj}^{0^I}} + \frac{\partial \widehat{y}_l^I}{\partial b_k}\frac{\partial b_k}{\partial q_k}\frac{\partial q_k}{\partial v_{kj}^{0^I}}\right]\end{aligned} \tag{2.28}$$

From Eq. (2.25), the following can be observed:

$$\frac{\partial p_k}{\partial v_{kj}^{0^R}} = z_j^R; \quad \frac{\partial p_k}{\partial v_{kj}^{0^I}} = -z_j^I; \quad \frac{\partial q_k}{\partial v_{kj}^{0^R}} = z_j^I; \text{ and} \frac{\partial q_k}{\partial v_{kj}^{0^I}} = z_j^R \tag{2.29}$$

Further, since the elementary transcendental function satisfies the essential properties of a complex-valued activation function, the Cauchy Riemann conditions can be used to derive the fully complex-valued BP algorithm. Hence,

$$\frac{\partial a_k}{\partial p_k} = \frac{\partial b_k}{\partial q_k}; \text{ and } \frac{\partial a_k}{\partial q_k} = -\frac{\partial b_k}{\partial p_k} \tag{2.30}$$

Thus, identifying the partial derivatives of Eqs. (2.27) and (2.28) from Eqs. (2.14), (2.29) and (2.30), we can obtain the gradient of the error function with respect to the input weights as:

$$\nabla v_{kj}^0 = -\delta_l\, \overline{v}_{lk}^1\, \overline{f}_a\left(\sum_j v_{kj}^0\, z_j\right)\overline{z}_j \tag{2.31}$$

$$\text{and} \Delta v_{kj}^0 = \eta_v\, \delta_l\left(\sum_{l=1}^{n} \overline{v}_{lk}^1 \overline{f}'_a\left(\sum_{j=1}^{m} v_{kj}^0 z_j\right)\right)\overline{z}_j;\ k = 1,\cdots,h,\ j = 1,\cdots,m; \tag{2.32}$$

It can be observed from the gradient based weight update rules derived in [1] and reproduced here, that the weight update rule for the fully complex-valued BP algorithm is the complex conjugate form of the real weight update formula. The network has m input neurons, h hidden neurons and n output neurons, all complex-valued signals. Hence, unlike SC-MLP network, FC-MLP network is compact and all the weights and neuron signals are complex-valued. Also, the network uses a fully complex-valued activation function, hence, it uses a complex-valued gradients that reflect the true gradients. Hence, they approximate the complex-valued signal better than SC-MLP networks, thereby, improving the phase approximation ability of FC-MLP network.

Apart from the ETF's, there are also other fully complex-valued activation functions discussed in the previous chapter [2]. One of them is

$$f_a(z) = f_a^R(z^R) + i f_a^I(z^I) \tag{2.33}$$

Such functions overcome the singularity issues of the ETF's. However as the choice of $f_a^R(.)$ and $f_a^I(.)$ is difficult, in this work,only the ETF's are considered as activation functions. It must be noted that irrespective of the activation function used, FC-MLP uses a complex-valued gradient descent based backpropagation learning algorithm. The performance of the algorithm is affected by the proper choice of the region and magnitude of weight initialization, the influence of learning rate parameter and the choice of number of neurons. In the next section, issues exisiting in the SC-MLP and FC-MLP networks are discussed.[4]

2.2 Issues in Fully Complex-valued Multi-Layer Perceptron Networks

In the MLP framework of the complex-valued network, two network configurations are available, viz., the SC-MLP network and the FC-MLP network. In general, both the SC-MLP network and FC-MLP network use a gradient based learning algorithm. The convergence of SC-MLP network depends largely on the learning rate parameter and the number of hidden neurons. On the other hand, the convergence of FC-MLP network depends on the magnitude and region of weight initialization, the learning rate parameter, the activation function used, the number of hidden neurons used and the error function/minimization function used in deriving the gradients. In this section, the convergence issues in the existing complex-valued MLP networks are discussed briefly.

[4] Note: The derivation of the complex-valued back propagation algorithm in this section assumes that the neurons in the output layer use a complex-valued activation function. In our study, we assume linear output neurons.

2.2.1 Split Complex-valued MLP

SC-MLP networks are similar in architecture to the real-valued MLPs and hence all real-valued activation functions are applicable here. These functions are bounded and analytic, thereby, the network has no singularity issues. However, SC-MLP networks suffer from their inability to approximate non-linear phase accurately [4]. This is because the gradients of a SC-MLP network are not truly complex-valued and do not represent the true complex-valued gradient. In other words, the derivatives in an SC-MLP network cannot fully exploit the correlation between the real and imaginary components of the weighted sum of input vectors [1]. Thereby, it may be inefficient in approximating the non-linear phase accurately.

2.2.2 Fully Complex-valued MLP

FC-MLP networks use fully complex-valued activation functions and complex-valued weights. The performance of a fully complex-valued MLP is affected by the complex-valued activation function used, the learning rate parameter, number of hidden neurons and the performance index used for minimization and to derive the gradients. In this section, we discuss in detail each of above mentioned issues.

2.2.2.1 Activation Function

According to Liouville's theorem [1], an entire and bounded function in the complex domain is a constant function. This makes the selection of activation function for an FC-MLP difficult. A set of ETF's which are bounded *almost everywhere* is a possible choice of activation functions. However, these functions have singularities associated with themselves and their derivatives. For example, the "*asinh*"activation function has two continuous branch cut singularities: one along the positive imaginary axis above i (inclusive) in quadrant I and another along the negative imaginary axis below $-i$ (inclusive) in quadrant III. During training if the weights are initialized such that the network operates in the singular region of the activation function, then the network fails to converge. Therefore, sufficient care has to be taken to ensure that the singular point of the activation functions or their singularities are not hit during the learning process.

The learning rate parameter: The learning rate parameter affects the convergence of the FC-MLP network. Too high a learning rate parameter results in faster convergence initially, but ends up with oscillating errors during learning. On the other hand, using a smaller value for the learning rate parameter results in a slower convergence also with a possibility of converging to local minima. Therefore, for an FC-MLP to successfully converge, a careful choice of the learning rate parameter becomes essential. Besides, for an FC-MLP, the learning rate parameter can be either real-valued or complex-valued. It will be an interesting future study to understand how a complex-valued learning rate interacts with the complex-valued gradients of the network.

Number of hidden neurons: In general, selection of the network size is a critical issue in a neural network model. One has to find a minimal architecture that accurately fits the true function described by the training data. A large network may accurately fit the training data, but may have poor generalization performance due to over-fitting. On the other hand, a smaller network requires lesser computational effort, but may not be able to approximate the given function. A heuristic procedure, similar to the one presented in [5] can be used to find the optimal network size. However, this is a time consuming procedure.

2.2.2.2 Performance Index

Another important aspect in a complex-valued back propagation algorithm is the selection of an appropriate performance index . (here the error function or the minimization criterion). In most of the algorithms presented in the literature, this product of the error and its conjugate (Euclidean norm), as given below, is used as an error function/performance measure for minimization.

$$E = \frac{1}{2n}(\mathbf{e}^H \mathbf{e}). \tag{2.34}$$

where $\mathbf{e} = \mathbf{e}_R + i\mathbf{e}_I = \mathbf{y} - \widehat{\mathbf{y}} = (\mathbf{y}_R - \widehat{\mathbf{y}}_R) + i(\mathbf{y}_I - \widehat{\mathbf{y}}_I)$, $\mathbf{e}^H$ is the complex-conjugate transpose of error, $\widehat{\mathbf{y}}$ is predicted output, $\mathbf{e}_R$ is the Real part of error and $\mathbf{e}_I$ is the Imaginary part of error.

From Eq. (2.34), one can see that the error function considers only the square magnitude error and does not consider the phase quantity of the error ($\mathbf{e}$) directly. Hence, the network evolved using such a error function may not tend to minimize the phase error directly and the network may not approximate the phase accurately, which is important for problems in the domain of telecommunication and image reconstruction. In general, for better approximation of complex-valued functions (i.e., both in magnitude and phase), one should find an appropriate complex-valued activation function and a suitable error function.

2.3 An Improved Fully Complex-valued Multi-Layer Perceptron (IC-MLP)

In the section 2.2.2.2, the issues in FC-MLP due to the mean squared error function were discussed. In this section, a new activation function for the FC-MLP is proposed and a new logarithmic error function/minimization criterion to aid better approximation of both the magnitude and phase of FC-MLP is developed. The network using logarithmic error function as the performance index and employing "*exp*" activation function at the output layer is termed as the Improved Complex-valued MLP (IC-MLP).

2.3.1 A New Activation Function: *exp*

The elementary transcendental functions available as activation functions for FC-MLP in the literature are sensitive to weight initializations and the learning rate parameter and suffer from issues arising due to the singularities in the finite region of the Complex plane. In this section, an exponential function is proposed as an activation function for the nonlinear processing of complex-valued data. The exponential function, $f_a(z) = exp(z)$, is entire in $\mathbb{C}$ since $f_a(z) = f_a'(z) = exp(z)$. The complex-valued exponential function has an essential singularity at $\pm\infty$. By restricting the weights of the network to a small ball of radius ($\mathbf{v}_r$) and the number of hidden neurons to a finite value, the bounded behavior in an FC-MLP network can be ensured.

Fig. 2.3 shows the magnitude and phase response plots for the "*exp*" activation function. From the plot, it can be seen that the activation function is continuous, an increasing function and is bounded in a bounded region of the Complex domain.

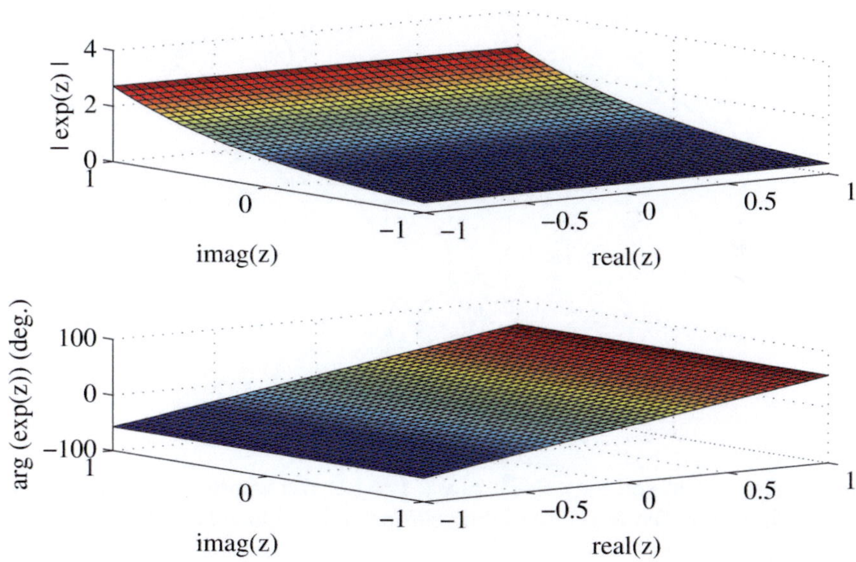

Fig. 2.3 Magnitude and Phase Response Plots of "*exp*" Activation Function

For illustration, the structure of the three layered IC-MLP network as shown in Fig. 2.4 is considered here.[1] As discussed earlier, let the number of neurons in the input, hidden and output layers are m, h and n respectively. Let $\mathbf{z} \in \mathbb{C}^m$ be the m dimensional complex-valued input to the network. The signals of each neuron in the input layer is denoted by z_k; $k = 1, 2, \ldots m$ where m is the dimension of the input.

[1] This can be easily extended with multiple hidden layers for multi-layer networks.

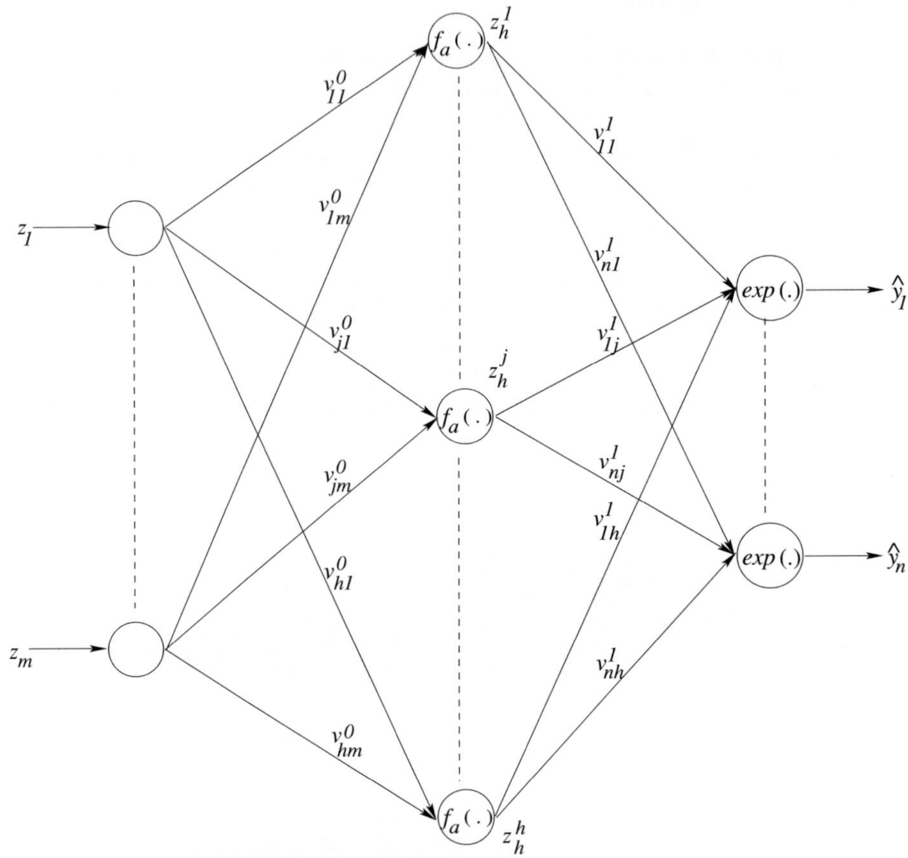

Fig. 2.4 Architecture of a IC-MLP Network

Let the targets be denoted by $\mathbf{y} \in \mathbb{C}^n = y_j;\ j = 1,2,...n$ where n is the dimension of the output vector. The response of the neurons in the hidden layer is given by:

$$z_h^k = f_a(\sum_j v_{kj}^0 z_j);\ \ k = 1,\cdots,h \tag{2.35}$$

It is notable that unlike the FC-MLP network, all the neurons in the output layer of IC-MLP network employ "*exp*" activation function. Therefore, the output of the l-th output neuron is given by:

$$\widehat{y}_l = exp\left(\sum_k v_{lk}^1 z_h^k\right);\ \ l = 1,\cdots,n \tag{2.36}$$

2.3.2 Logarithmic Performance Index

In the literature, mean squared error function is often used aas the minimization criterion. For complex-valued signals, the squared error represents only the magnitude of error explicitly and does not include the phase error directly. Hence, a function that includes both the magnitude and phase errors directly becomes essential. Therefore, we propose a new error function which includes both the magnitude and phase errors directly and a fully complex-valued error gradient. The new error function is defined as

$$E_{new} = \frac{1}{2}\left[log\left[\frac{\mathbf{y}}{\widehat{\mathbf{y}}}\right]\overline{log\left[\frac{\mathbf{y}}{\widehat{\mathbf{y}}}\right]}\right]; \quad \widehat{\mathbf{y}} = [\widehat{y}_1, \cdots, \widehat{y}_l, \cdots, \widehat{y}_n]^T \tag{2.37}$$

where $\overline{log\left[\frac{\mathbf{y}}{\widehat{\mathbf{y}}}\right]}$ is the complex-conjugate of $log\left[\frac{\mathbf{y}}{\widehat{\mathbf{y}}}\right]$. This is different from eq. (2.34) (section 2.2.2) in that eq. (2.34) includes only the magnitude of the complex-valued signal explicitly, while eq. (2.37) includes both the error in magnitude and phase, explicitly. The above equation can be written as

$$E_{new} = \frac{1}{2}\left[log\left[\frac{||\,\mathbf{y}\,||}{||\,\widehat{\mathbf{y}}\,||}\right]^2 + [\angle(\mathbf{y}) - \angle(\widehat{\mathbf{y}})]^2\right] \tag{2.38}$$

Constants k_1 and k_2 can be included as weighting factors for the magnitude error and phase error, respectively to accelerate the convergence of the magnitude/phase component of the output. With the introduction of constants k_1 and k_2, eq.(2.38) becomes

$$E_{new} = \frac{1}{2}\left[k_1 log\left[\frac{||\,\mathbf{y}\,||}{||\,\widehat{\mathbf{y}}\,||}\right]^2 + k_2[arg(\mathbf{y}) - arg(\widehat{\mathbf{y}})]^2\right] \tag{2.39}$$

From the above equation, one can see that the performance index (E_{new}) approaches zero when $\widehat{\mathbf{y}}$ approaches $\mathbf{y}$. Here, the targets are all scaled into first and fourth quadrants. k_1 and k_2 are weights applied to the logarithmic magnitude error and phase error respectively. It should be noted here that while any of the ETF's or "*exp*" activation function may be employed at the hidden layer, all the neurons in the output layer employ "*exp*" activation function. The "*exp*" activation function at the output layer helps to avoid ($\frac{1}{\widehat{\mathbf{y}}}$) term that is a part of the derivative of logarithmic error in the backward computation. Thus, the output of neurons in the hidden layer are computed as

$$z_h^j = f_a\left[\sum_{k=1}^{m}(v_{jk}^0 z_j)\right], \quad j = 1, 2, \ldots.h, \tag{2.40}$$

where $f_a : \mathbb{C}^m \rightarrow \mathbb{C}$,is any ETF or the "*exp*" activation functions and v^0_{jk} is the complex-valued weight between j^{th} neuron in hidden layer and k^{th} neuron in input layer. The output of the network is computed as

$$\widehat{y}_j = exp\left[\sum_{k=1}^{h}(v^1_{jk} z^k_h)\right] \quad , \quad j = 1,2,....n, \tag{2.41}$$

2.3.3 *Learning Algorithm*

The derivation of the complex-valued back propagation algorithm based on the logarithmic error function is similar to the complex-valued back propagation algorithm derived in section 2.1.2, except for the term δ_l, the error at the output layer. Here,

$$\delta_l = \left(k_1\ log\left(\frac{|\ y_l\ |}{|\ \widehat{y}_l\ |}\right) + ik_2\angle y_l - \angle \widehat{y}_l\right) \tag{2.42}$$

For the update of v^1_{lk}, the weight connecting the lth output neuron and the kth hidden neuron, . Hence,

$$\Delta v^1_{lk} = \eta\ \overline{z^k_h}\ \left(k_1\ log\left(\frac{|\ y_l\ |}{|\ \widehat{y}_l\ |}\right) + ik_2\angle y_l - \angle \widehat{y}_l\right);\ l = 1,\cdots,n \tag{2.43}$$

and the update of v^0_{kj}, the weight connecting the kth hidden neuron and the jth input neuron is given by

$$\Delta v^0_{jk} = \eta\left(\sum_{l=1}^{n}\overline{v}^1_{lk}\overline{f}'_a(\sum_{j=1}^{m} v^0_{kj} z_j)\right)\left(k_1 log\left(\frac{|\ y_l\ |}{|\ \widehat{y}_l\ |}\right) + ik_2\angle y_l - \angle \widehat{y}_l\right)\overline{z}_j; l = 1,\cdots n \tag{2.44}$$

As the logarithmic error function includes both the magnitude and phase error explicitly and tends to minimize both simultaneously, the algorithm can improve the magnitude and phase approximation performance of FC-MLP.

2.4 Summary

In this chapter, different types of complex-valued MLP networks, viz., SC-MLP networks and FC-MLP networks were presented in detail. The issues in SC-MLP and FC-MLP were highlighted and the influence of the singularities of the activation functions on the performance of FC-MLP was discussed. It was also noted that the mean-squared error criterion that is usually used to minimize errors is not an appropriate performance measure in complex-valued networks. To overcome these issues, we have proposed the $exp(.)$ activation function with its singularity at $\pm\infty$ as an activation function for complex-valued MLP networks and a logarithmic error function which minimized both the magnitude and phase of the error simultaneously. We have also developed an IC-MLP with the *exp* activation function at the output layer and the logarithmic error function.

References

1. Kim, T., Adali, T.: Fully complex multi-layer perceptron network for nonlinear signal processing. Journal of VLSI Signal Processing 32(1/2), 29–43 (2002)
2. You, C., Hong, D.: Nonlinear blind equalization schemes using complex-valued multilayer feedforward neural networks. IEEE Transactions on Neural Networks 9(6), 1442–1455 (1998)
3. Georgiou, G.M., Koutsougeras, C.: Complex domain backpropagation. IEEE Transactions on Circuits and Systems-II: Analog and Digital Signal Processing 39(5), 330–334 (1992)
4. Kim, T., Adali, T.: Approximation by fully complex multi-layer perceptrons. Neural Computation 15(7), 1641–1666 (2003)
5. Suresh, S., Omkar, S.N., Mani, V., Prakash, T.N.G.: Lift coefficient prediction at high angle of attack using recurrent neural network. Aerospace Science and Technology 7(8), 595–602 (2003)

Chapter 3
A Fully Complex-valued Radial Basis Function Network and Its Learning Algorithm

Radial basis function networks are the most popular neural network architecture due its simpler structure and better approximation ability owing to the localization property of the Gaussian function. in this chapter, we study complex-valued RBF networks and their learning algorithms. First, we present a complex-valued RBF network which is a direct extension of the real-valued RBF network. CRBF network is a single hidden layer network which computes the output of the network as a linear combination of the hidden neuron outputs. The hidden neurons map the input space to a nonlinear space using a radially symmetric basis function ('Gaussian function'). The CRBF network uses the real-valued Gaussian activation function that maps $\mathbb{C}^m \to \mathbb{R}$ as a basis function. Hence, despite the centers of the Gaussian neurons and the output weights being complex-valued, the response of the hidden neurons remains real-valued. Moreover, during the backward learning, the real part of the error is used to update the real part of the network parameters and the imaginary part of the error is used to update the imaginary part of the network parameters. Hence, the gradients used are not a true representation of the gradient of the target function and hence, will not approximate phase accurately, as explained in[1]. Hence, there is a need to develop a fully complex Gaussian like symmetric function, which maps $\mathbb{C}^m \to \mathbb{C}$. In this chapter, we address this issue by proposing a fully complex-valued RBF network with "*sech*" function as the basis function. The "*sech*" activation function is symmetric about the real axis and maps $\mathbb{C}^m \to \mathbb{C}$. The gradient descent based learning algorithm of the FC-RBF network with "*sech*" is also derived for tuning the free parameters of the network.

In most of the batch learning algorithms available in the literature, it is implicitly assumed that the training samples are uniformly distributed in the input space. This assumption is not always true especially in real-world problems and hence affects the generalization ability of the trained network. Therefore, we need to enable a scheme for selective participation of the samples in the learning process in order to improve the generalization ability of the network. Since most of the machine learning paradigms are inspired from the principles of human learning, in this chapter, we emulate the principles of human meta-cognition in a neural network framework

S. Suresh et al.: Supervised Learning with Complex-valued Neural Networks, SCI 421, pp. 49–71.
springerlink.com

by developing a 'Meta-cognitive Fully Complex-valued Radial Basis Function (Mc-FCRBF) Network' . Mc-FCRBF is developed based on the Nelson and Narens [2] model of meta-cognition in the context of human learning . This chapter describes the architecture and learning algorithm of Mc-FCRBF in detail.

3.1 Complex-valued RBF Networks

In this section, a discussion on different existing complex-valued RBF networks is presented first. Complex-valued RBF networks [3, 4] are the extensions of real-valued RBF networks for operating on complex-valued signals. As described in chapter 2, let $\{(\mathbf{z}_1, \mathbf{y}_1), (\mathbf{z}_2, \mathbf{y}_2), \cdots, (\mathbf{z}_t, \mathbf{y}_t), \cdots (\mathbf{z}_N, \mathbf{y}_N)\}$, where $\mathbf{z}_t \in \mathbb{C}^m$; $t = 1, 2, \cdots N$, where N is the total number of observations in the training data set and $\mathbf{y}_t \in \mathbb{C}^n$ be the observation/training data set. For notational convenience, the subscript t is being dropped in future discussions.

CRBF network has m input neurons, h hidden neurons, and n output neurons. The input neurons are linear and just relay the inputs/features to the hidden layer. The neurons in the hidden layer of CRBF employ Gaussian activation function ($f_g(.)$). Each hidden neuron estimates the localized response for a given input. Finally, the n neurons in the output layer of CRBF obtain the weighted sum of the hidden layer responses to generate the n-dimensional complex-valued output ($\widehat{\mathbf{y}} = [\widehat{y}_1\ \widehat{y}_2 \cdots \widehat{y}_n]$).

Fig. 3.1 shows the architecture for a complex-valued RBF network. It can be observed from figure 3.1 that the CRBF network maps $\mathbb{C}^m \rightarrow \mathbb{C}^n$. However, at the hidden neurons, Gaussian function is employed as the activation function. The response of the jth hidden neuron is given by eq. (3.1)

$$y_h^j = \phi(\mathbf{z}, \mathbf{c}_j, \sigma_j) = \exp\left(-\frac{(\mathbf{z}-\mathbf{c}_j)^H(\mathbf{z}-\mathbf{c}_j)}{2\sigma_j^2}\right); \ j = 1, 2, \cdots h; \qquad (3.1)$$

where, H denotes the Hermitian of the complex-valued signal, $\mathbf{c}_j \in \mathbb{C}^m$ is the complex-valued center of the jth hidden neuron employing the Gaussian function. $\sigma_j \in \mathbb{R}$ is width of the jth hidden neuron employing the Gaussian function. It can be observed from eqn. (3.1) that the activation function considers the product of the hermitian of a complex-valued vector $(\mathbf{z}-\mathbf{c}_j)$ and the vector itself. Hence, despite the center of the radial basis function and the output weights being complex-valued, the output of the hidden neurons are real.

The output of the CRBF network, $\widehat{y}_i$ is given by eq. (3.2).

$$\widehat{y}_k = \sum_{j=1}^{h} v_{kj} y_h^j, \quad k = 1, 2, \ldots. n. \qquad (3.2)$$

where v_{kj} is the complex-valued weight connecting the jth hidden neuron and the kth output neuron. Now, we present the gradient based update rules for the CRBF network.

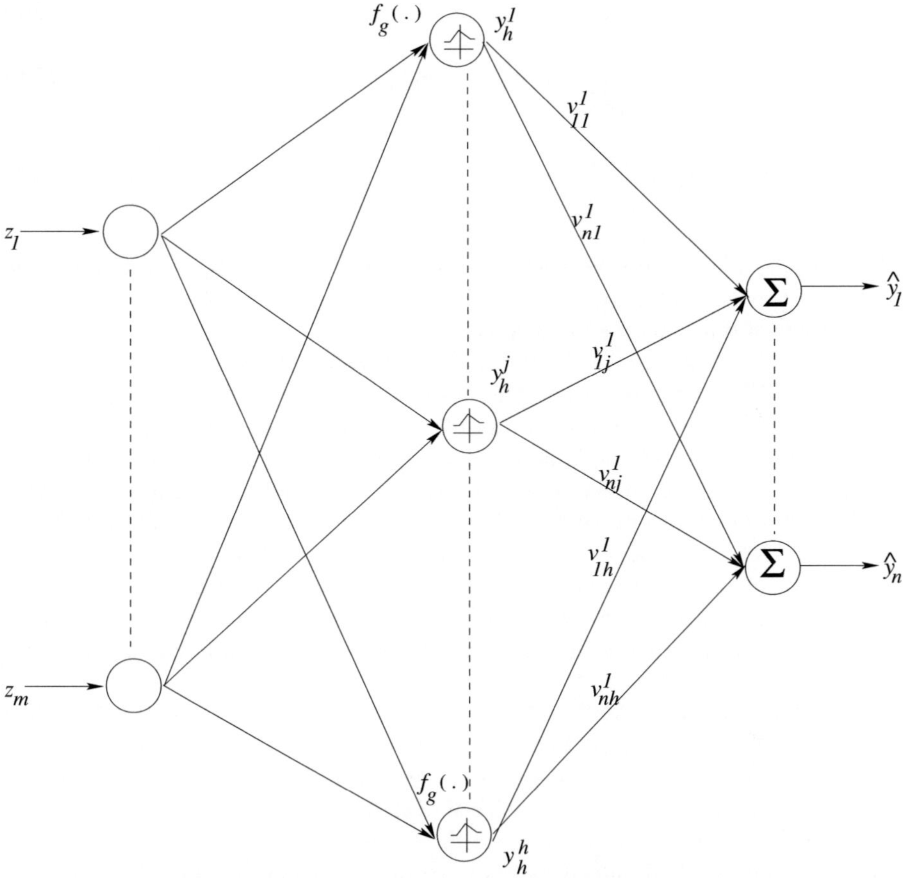

Fig. 3.1 Architecture of a Complex-valued RBF Network

The error for the CRBF network is given by:

$$\mathbf{e} = \mathbf{e}_R + i\mathbf{e}_I = \mathbf{y} - \widehat{\mathbf{y}} \tag{3.3}$$

Let $\mathbf{e}_R$ and $\mathbf{e}_I$ be the real and imaginary components of the complex-valued error $\mathbf{e}$. In [5], a stochastic gradient training algorithm that adapts all the free parameters ($\mathbf{c}$, $\boldsymbol{\sigma}$ and V) of the network has been presented, where, the update rules are derived based on the mean squared error as given by:

$$E = \frac{1}{2}\sum_{k=1}^{n} e_k \bar{e}_k \tag{3.4}$$

The update rules for the free parameters of the network are given by Eqs. (3.5), (3.6) and (3.7).

$$\Delta v_{kj} = \mu_v \, e_k \, y_h^j; \;\; k = 1,2,\cdots n; \;\; j = 1,2,\cdots,h \tag{3.5}$$

$$\Delta \sigma_j = \mu_\sigma \, y_h^j \left[\sum_{k=1}^{n} \left(v_{kj}^R \, e_k^R + v_{kj}^I \, e_k^I \right) \right] \cdot \frac{\| \mathbf{z} - \mathbf{c}_j \|^2}{\sigma_j^3} \tag{3.6}$$

$$\Delta \mathbf{c}_j = \mu_c \, y_h^j \left[\frac{\sum_{k=1}^{n} \left[v_{kj}^R \, e_k^R \, re(\mathbf{z} - \mathbf{c_j}) + i v_{kj}^I \, e_k^I \, im(\mathbf{z} - \mathbf{c_j}) \right]}{\sigma_j^2} \right] \tag{3.7}$$

where μ_v, μ_σ and μ_c are the learning rate parameters for weight, width and the centers of Gaussian function, respectively. v_{kj}^R and v_{kj}^I are the Real and Imaginary components of this weight v_{kj} respectively.

From eq. (3.1) and eq. (3.2) one can note that though the centers and weights of the CRBF network are all complex-valued, the response of the hidden neuron is real-valued, and hence, does not transmit the complex-valued input signal to the output neurons fully. This affects the phase approximation ability of the network. Besides, it may also be observed from eqs. (3.5)-(3.7) that the gradients use the Real and Imaginary components of the complex-valued error and the output weights to update the free parameters of the network. This does not capture the correlation between the Real-Imaginary components of the complex-valued gradient. Hence, the gradient thus derived, is not a true representation of the true complex-valued gradient. This is also due to the fact that the responses at the hidden neurons are real-valued.

In the literature, this CRBF network has been used in several applications like channel equalization, adaptive beam forming [4, 5, 6] etc. Also, several real-valued batch and sequential learning algorithms have been extended to the Complex domain in this framework using a Gaussian activation function. For example, complex-valued extreme learning machine [7] is a direct extension of the real-valued extreme learning machine. Real-valued extreme learning machine assumes random real-valued input weights and estimates the output weights analytically. Similarly, the C-ELM algorithm with Gaussian activation function also assumes random complex-valued centers and estimates the complex-valued output weights analytically. The least-squared solution of V, of the linear system $\mathbf{y}_h V = Y$ with *minimum norm* of output weights V, which usually tend to have good generalization performance [7]. If $\mathbf{y}_h \in \mathbb{R}^h$ (RBF hidden neurons employing Gaussian activation function) is the vector of hidden neuron outputs, the output weights ($V \in \mathbb{C}^{n \times h}$) are estimated using

$$V = \mathbf{y}_h^+ Y \tag{3.8}$$

where $\mathbf{y}_h^+$ is the Moore-Penrose generalized inverse of the complex-valued vector $\mathbf{y}_h$ and $Y = [\mathbf{y}_1, \; \mathbf{y}_2, \; \cdots \; \mathbf{y}_N]$. Similarly, the complex-valued minimal resource allocation network [8] and the complex-valued growing and pruning RBF network [9] are direct extensions of the real-valued sequential learning algorithms, the minimal resource allocation network [10, 11] and growing and pruning network-radial basis function networks [12], respectively. As these algorithms have been extended to the complex domain in the framework of the CRBF network, the gradients used in these algorithms also use the Real and Imaginary components of the complex-valued signal. Hence, the correlations between the Real-Imaginary components is lost and hence, the gradients are not true representations of the complex-valued gradients. Hence, these algorithms are also ineffective in approximating phase accurately. In the next section, the issues related to the CRBF networks are summarised briefly.

3.2 Factors Influencing the Performance of Complex-valued RBF Networks

Existing complex-valued RBF networks suffer from the following issues:

- ***Activation function:*** As shown in the section 3.1, the Gaussian function is used as the activation function and the response of the hidden neuron is given by:

$$y_h^j = \phi\left(\frac{\mathbf{z}-\mathbf{c}_j}{\sigma_j}\right) = \exp\left(-\frac{(\mathbf{z}-\mathbf{c}_j)^H(\mathbf{z}-\mathbf{c}_j)}{2\sigma_j^2}\right); \tag{3.9}$$

  This function maps $\mathbb{C}^m \to \mathbb{R}$, due to the presence of the Hermitian operator in the activation function. Hence, despite the inputs, weights, and centers being complex-valued, the response at the hidden node is still real-valued. This affects the phase approximation performance of the network. Also, as seen from eqs. (3.5)-(3.7), the gradients that are used in updating the free parameters of the network are not fully complex-valued. Instead, they use the real-valued Real and Imaginary components of the complex-valued error and weights. Thus, as the correlation between the Real and Imaginary components of the weights and errors (and their products) is lost, this representation does not reflect the true complex-valued gradient. As these gradients lack vital data about the complex-valued error signal, the approximation of a complex-valued function using the network is inefficient.
- ***Number of hidden neurons:*** The selection of the number of hidden neurons for the network is done by a heuristic procedure that is similar to the procedure presented in [13] for real-valued networks. Accordingly, the search begins with a smaller number of hidden neurons, and observing the performances for different network sizes. For a given initialization and learning rate parameters, the optimal number of neurons is chosen as the network size at which both the training and generalization performance of the network is better than other choices of network size. Selection of fewer neurons result in insufficient neurons and hence, poor approximation performance. Larger network structure results in poor generalization performance and also adds to the computational cost.
- ***Initialization:*** The initialization of centers and weights affects the performance of the CRBF network significantly. If the centers are located either sparsely or closely, the input space is not efficiently projected onto the Gaussian space, resulting in poor approximation/classification performance.

In the next section, a fully complex-valued RBF network with a fully complex-valued activation function "*sech*"whose magnitude response is similar to that of the real-valued Gaussian activation function is presented. The activation function maps $\mathbb{C}^m \rightarrow \mathbb{C}$. Hence, it ensures that the gradients are also fully complex-valued, thus, improving the phase approximation performance. The issues due to the network structure and initialization are overcome with an K-Means Clustering (KMC) algorithm for the selection of neurons, initial centers and weights of the FC-RBF network.

3.3 A Fully Complex-valued RBF Network (FC-RBF)

In this section, we propose a fully complex-valued RBF network with a fully complex-valued activation function. As will be seen, the activation function is symmetric and has a magnitude response similar to that of the real-valued Gaussian activation function.

3.3.1 Network Architecture

Similar to the CRBF network, the FC-RBF network has one input layer, one hidden layer and one output layer. The architecture of FC-RBF network is given in Fig 3.2. As can be observed from the figure, the network has m input neurons, h hidden neurons and n output neurons. The weights between the input layer and the hidden layer are unity. At the hidden layer, the localized response of the inputs are computed with the non-linear fully complex-valued activation function for the neuron "*sech*" function. The activation function used in the hidden neurons is described in eq. (3.10).

$$z_h^j = f_a(z) = sech(\boldsymbol{\sigma}_j^T.(\mathbf{z}-\mathbf{c}_j)) \ , \quad j = 1,2,....,h. \tag{3.10}$$

where $\boldsymbol{\sigma}_j \in \mathbb{C}^m$ is the complex-valued scaling factor[1] and $\mathbf{c}_j \in \mathbb{C}^m$ the center of the jth hidden neuron. h is the number of hidden neurons. The scaling factor $\boldsymbol{\sigma}_j$ plays a role similar to the deviation σ in the real-valued Gaussian function. The output of the network $\widehat{y}_k$ is given by eq. (3.11).

$$\widehat{y}_k = \sum_{j=1}^{h} v_{kj} \, z_h^j, \quad k = 1,2,\cdots n. \tag{3.11}$$

where the $v_{kj} \in \mathbb{C}$ are the complex-valued output weights.

3.3.2 The Activation Function

The network uses "*sech*" as a basis for a fully complex-valued activation function and the functional unit of the hidden neuron using "*sech*" is the form shown in eq.

[1] $\boldsymbol{\sigma}_j$ is a complex-valued vector.

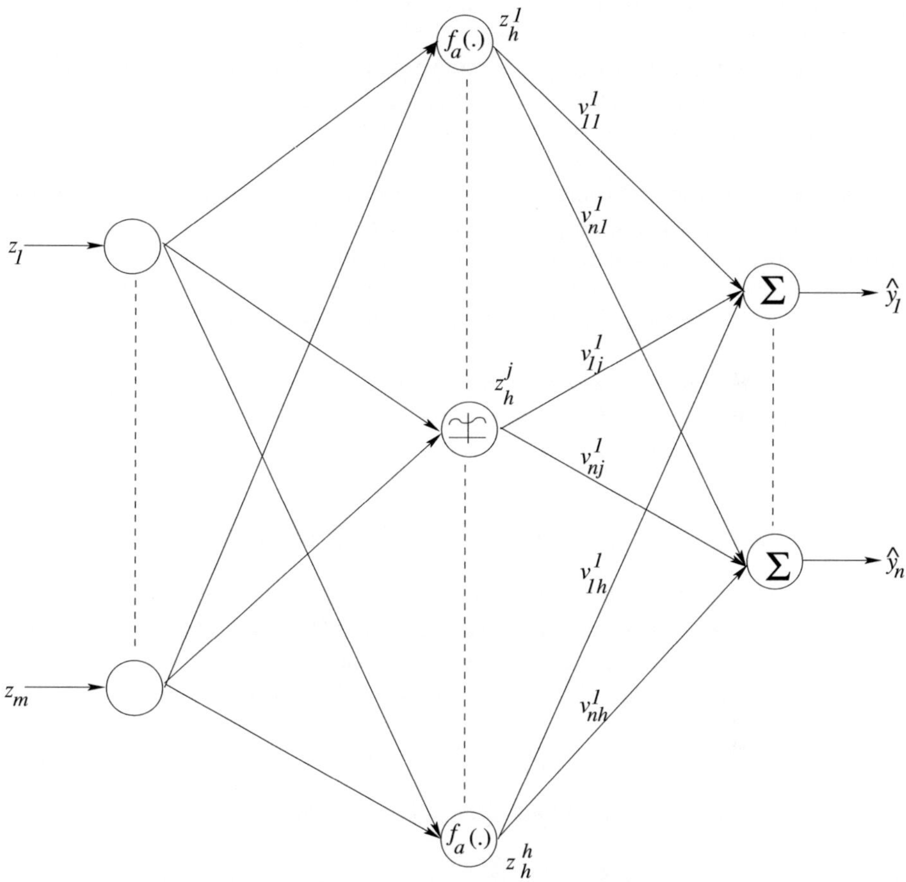

Fig. 3.2 Architecture of a Fully Complex-valued RBF Network

(3.12). The Gaussian like response of the "*sech*" function can be seen from Fig. 3.3(a).

$$z_h^j = sech(\boldsymbol{\sigma}_j^T . (\mathbf{z} - \mathbf{c}_j)); \; j = 1, 2, \cdots h \tag{3.12}$$

The activation function "*sech*" satisfies the desired properties for a complex-valued activation function stated in [14], namely,

> *In a bounded domain of complex plane $\mathbb{C}$, a fully complex nonlinear activation function $f(z)$ needs to be analytic and bounded almost everywhere.*

"*sech*" has periodic isolated singularities at $(1/2 + n)\pi\ i$ where $n \in \mathbb{N}$. Therefore, it is analytic and bounded *almost everywhere* in a bounded domain of the Complex plane. The proof for the approximation capability of "*sech*" activation, with isolated singularities, is straightforward from [15], i.e., the isolated singularity of "*sech*" can provide uniformly converging approximation in the deleted annulus of the singularity in the region of convergence defined by the Laurent series. Figs. 3.3(a) and 3.3(b) give the magnitude and phase response plots of this function over the complex plane. It can be seen from the figures that the magnitude response of the fully complex-valued "*sech*" activation function is similar to that of the Gaussian activation function, i.e., the magnitude of the topological neighborhood of z_h^j decreases monotonically with increasing difference $\mathbf{z} - \mathbf{c}_j$, the magnitude being the highest when $\mathbf{z} - \mathbf{c}_j = 0$. Also, the phase response of the function (Fig. 3.3(b)) is close to zero when both the real and imaginary parts of the complex-valued signal $\boldsymbol{\sigma}_j^T.(\mathbf{z} - \mathbf{c}_j)$ are zero. Thus the activation function response is the highest when the inputs are located closer to the center, making the function a good choice for a radial basis function.

In the next section, the gradient descent based learning algorithm for the FC-RBF network is presented in detail. The gradient descent based parameter updates for the free parameters of the network are derived.

3.4 Learning Algorithm for the FC-RBF Network

The complex-valued "*sech*" in section 3.3 satisfies the desirable properties of complex-valued activation functions presented in [14]. It is analytic and bounded almost everywhere. Therefore, the Cauchy-Riemann equations defined in eq. (2.11) can be used to simplify the fully complex-valued RBF algorithm. i.e.,

$$f_a'(z) = f_a^R = -i f_a^I \tag{3.13}$$

The sum-squared error at the output layer can be written as

$$E = \frac{1}{2}\mathbf{e}^H\,\mathbf{e} = \sum_k e_k\,\overline{e}_k; \text{ where } e_k = y_k - \widehat{y}_k;\; k = 1,2,\cdots n, \tag{3.14}$$

where $\overline{e}_k$ is the conjugate of error of the k-th output neuron.

With h neurons in the hidden layer, let $z_h^j = a_j + ib_j$ be the response of the jth hidden neuron,

$$\begin{aligned}\widehat{y}_k &= \sum_{j=1}^{h} (v_{kj} z_h^j) = \sum_{j=1}^{h} (v_{kj}^R + i\,v_{kj}^I)(a_j + jb_j)\\ &= \sum_{j=1}^{h} (v_{kj}^R\,a_j - v_{kj}^I\,b_j) + i\,(v_{kj}^R\,b_j + v_{kj}^I\,a_j);\; k = 1,2,\cdots n\end{aligned} \tag{3.15}$$

where the superscripts R and I indicate the real and imaginary components, respectively.

The output weight update rule requires the computation of the gradient $\frac{\partial E}{\partial v_{kj}}$. Since the cost function is a real-valued function of a complex-valued variable, the

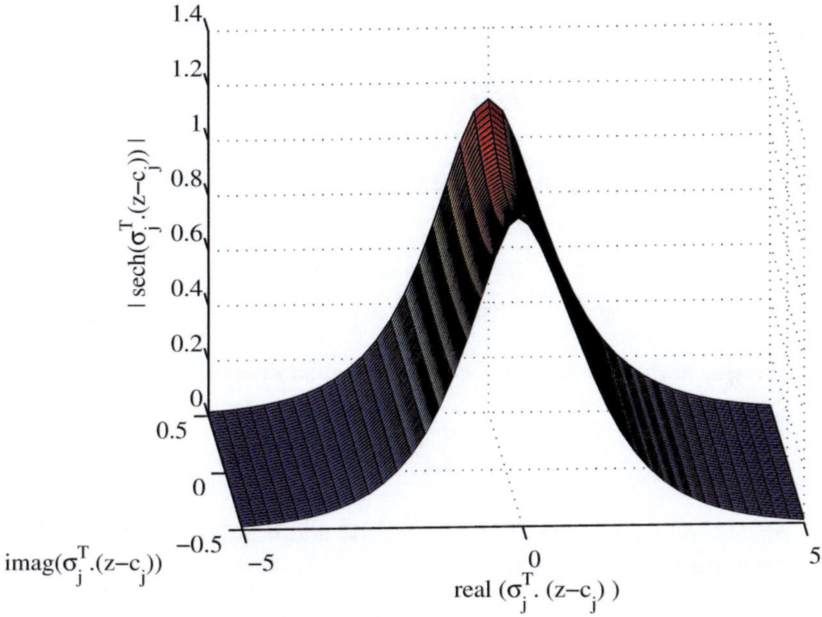

(a) Magnitude Response Plot of "*sech*"

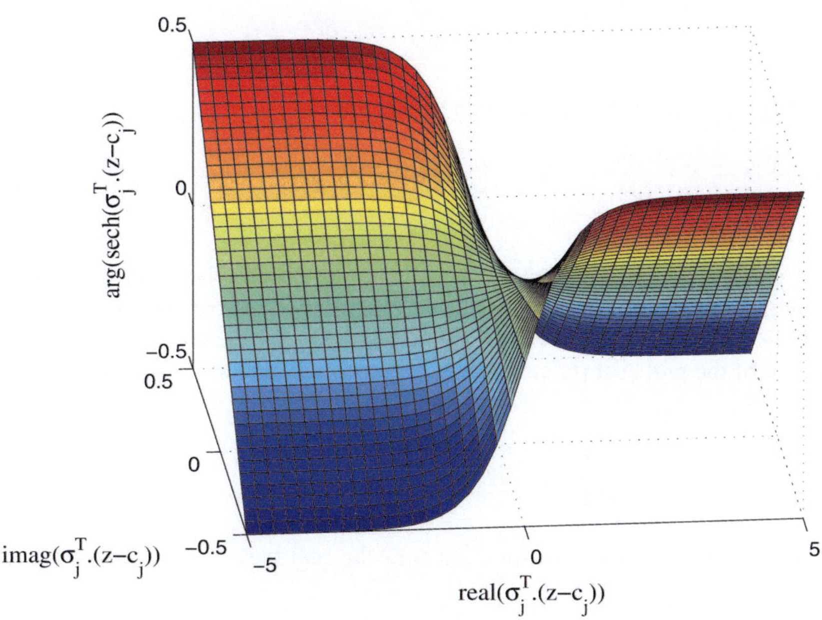

(b) Phase Response Plot of "*sech*"

Fig. 3.3 Responses of "*sech*" (a) Magnitude response (b) Phase response gradient of the error function with respect to the Real and Imaginary components of the v_{kj} are given by

$$\frac{\partial E}{\partial v_{kj}} = \nabla v_{kj} E = \frac{\partial E}{\partial v_{kj}^R} + i\,\frac{\partial E}{\partial v_{kj}^I} \tag{3.16}$$

Using the chain rule, the derivative of the cost function with respect to (w.r.t) the real part of v_{kj} is given by:

$$\frac{\partial E}{\partial v_{kj}^R} = \frac{\partial E}{\partial \widehat{y}_k^R}\frac{\partial \widehat{y}_k^R}{\partial v_{kj}^R} + \frac{\partial E}{\partial \widehat{y}_k^I}\frac{\partial \widehat{y}_k^I}{\partial v_{kj}^R} \tag{3.17}$$

and the derivative of the cost function w.r.t the imaginary part of v_{kj} is given by

$$\frac{\partial E}{\partial v_{kj}^I} = \frac{\partial E}{\partial \widehat{y}_k^R}\frac{\partial \widehat{y}_k^R}{\partial v_{kj}^I} + \frac{\partial E}{\partial \widehat{y}_k^I}\frac{\partial \widehat{y}_k^I}{\partial v_{kj}^I} \tag{3.18}$$

Defining $\delta_k^R \equiv -\partial E/\partial \widehat{y}_k^R$ and $\delta_k^I \equiv -\partial E/\partial \widehat{y}_k^I$ and using the following partial derivatives obtained from eq. (3.15),

$$\begin{aligned} &\frac{\partial \widehat{y}_k^R}{\partial v_{kj}^R} = a_j;\ \frac{\partial \widehat{y}_k^I}{\partial v_{kj}^R} = b_j;\ \frac{\partial \widehat{y}_k^R}{\partial v_{kj}^I} = -b_j;\ \frac{\partial \widehat{y}_k^I}{\partial v_{kj}^I} = a_j \\ &\frac{\partial E}{\partial v_{kj}^R} = -\delta_k^R a_j - \delta_k^I b_j;\ \frac{\partial E}{\partial v_{kj}^I} = -\delta_k^R(-b_j) - \delta_k^I a_j \end{aligned} \tag{3.19}$$

From (3.19) the gradient of the error can be written as

$$\nabla v_{kj} E = (-\delta_k^R a_j - \delta_k^I b_j) + i\,(\delta_k^R b_j - \delta_k^I a_j) = -(\delta_k^R + i\delta_k^I)\,\bar{z}_h^j \tag{3.20}$$

$$\triangle v_{kj} = \eta_{\mathbf{v}}\,\bar{z}_h^j\,\delta_k \tag{3.21}$$

where $\bar{z}_h^j$ denotes the complex-conjugate of z_h^j and $\eta_{\mathbf{v}}$ is the learning rate for the output weight and can be either real-valued, imaginary-valued or complex-valued.

Similarly, letting $p_j + jq_j = \boldsymbol{\sigma}_j^T.(\mathbf{z} - \mathbf{c}_j)$, the update for the weights $\boldsymbol{\sigma}_j$ requires the gradient of the real cost function with respect to the real and imaginary components of $\boldsymbol{\sigma}_j$.

$$\frac{\partial E}{\partial \boldsymbol{\sigma}_j} = \nabla \boldsymbol{\sigma}_j E = \frac{\partial E}{\partial \boldsymbol{\sigma}_j^R} + i\,\frac{\partial E}{\partial \boldsymbol{\sigma}_j^I} \tag{3.22}$$

where the derivative of the cost function w.r.t the real part of $\boldsymbol{\sigma}_j$ is given by eq. (3.23).

$$\frac{\partial E}{\partial \boldsymbol{\sigma}_j^R} = \frac{\partial E}{\partial \widehat{y}_k^R} \left[\frac{\partial \widehat{y}_k^R}{\partial a_j} \frac{\partial a_j}{\partial p_j} \frac{\partial p_j}{\partial \boldsymbol{\sigma}_j^R} + \frac{\partial \widehat{y}_k^R}{\partial a_j} \frac{\partial a_j}{\partial q_j} \frac{\partial q_j}{\partial \boldsymbol{\sigma}_j^R} + \frac{\partial \widehat{y}_k^R}{\partial b_j} \frac{\partial b_j}{\partial p_j} \frac{\partial p_j}{\partial \boldsymbol{\sigma}_j^R} + \frac{\partial \widehat{y}_k^R}{\partial b_j} \frac{\partial b_j}{\partial q_j} \frac{\partial q_j}{\partial \boldsymbol{\sigma}_j^R} \right]$$
$$+ \frac{\partial E}{\partial \widehat{y}_k^I} \left[\frac{\partial \widehat{y}_k^I}{\partial a_j} \frac{\partial a_j}{\partial p_j} \frac{\partial p_j}{\partial \boldsymbol{\sigma}_j^R} + \frac{\partial \widehat{y}_k^I}{\partial a_j} \frac{\partial a_j}{\partial q_j} \frac{\partial q_j}{\partial \boldsymbol{\sigma}_j^R} + \frac{\partial \widehat{y}_k^I}{\partial b_j} \frac{\partial b_j}{\partial p_j} \frac{\partial p_j}{\partial \boldsymbol{\sigma}_j^R} + \frac{\partial \widehat{y}_k^I}{\partial b_j} \frac{\partial b_j}{\partial q_j} \frac{\partial q_j}{\partial \boldsymbol{\sigma}_j^R} \right] \tag{3.23}$$

and the derivative of the cost function w.r.t the imaginary component of σ_j is given by eq. (3.24)

$$\frac{\partial E}{\partial \boldsymbol{\sigma}_j^I} = \frac{\partial E}{\partial \widehat{y}_k^R} \left[\frac{\partial \widehat{y}_k^R}{\partial a_j} \frac{\partial a_j}{\partial p_j} \frac{\partial p_j}{\partial \boldsymbol{\sigma}_j^I} + \frac{\partial \widehat{y}_k^R}{\partial a_j} \frac{\partial a_j}{\partial q_j} \frac{\partial q_j}{\partial \boldsymbol{\sigma}_j^I} + \frac{\partial \widehat{y}_k^R}{\partial b_j} \frac{\partial b_j}{\partial p_j} \frac{\partial p_j}{\partial \boldsymbol{\sigma}_j^I} + \frac{\partial \widehat{y}_k^R}{\partial b_j} \frac{\partial b_j}{\partial q_j} \frac{\partial q_j}{\partial \boldsymbol{\sigma}_j^I} \right] +$$
$$\frac{\partial E}{\partial \widehat{y}_k^I} \left[\frac{\partial \widehat{y}_k^I}{\partial a_j} \frac{\partial a_j}{\partial p_j} \frac{\partial p_j}{\partial \boldsymbol{\sigma}_j^I} + \frac{\partial \widehat{y}_k^I}{\partial a_j} \frac{\partial a_j}{\partial q_j} \frac{\partial q_j}{\partial \boldsymbol{\sigma}_j^I} + \frac{\partial \widehat{y}_k^I}{\partial b_j} \frac{\partial b_j}{\partial p_j} \frac{\partial p_j}{\partial \boldsymbol{\sigma}_j^I} + \frac{\partial \widehat{y}_k^I}{\partial b_j} \frac{\partial b_j}{\partial q_j} \frac{\partial q_j}{\partial \boldsymbol{\sigma}_j^I} \right] \tag{3.24}$$

Identifying the partial derivatives of eq.(3.23) and (3.24) from eq. (3.13), (3.15) and (3.19), leads to

$$\frac{\partial E}{\partial \boldsymbol{\sigma}_j} = \delta_k \, \overline{v}_{kj} \, \overline{f_a}'(\boldsymbol{\sigma}_j^T.(\mathbf{z}-\mathbf{c}_j)) \; (\overline{\mathbf{z}-\mathbf{c}_j}) \tag{3.25}$$

where, $\overline{f_a}'$ is the conjugate of the derivative of the function f_a, and hence

$$\triangle \boldsymbol{\sigma}_j = \eta_{\boldsymbol{\sigma}} \; \delta_k \, \overline{v}_{kj} \, \overline{f_a}'(\boldsymbol{\sigma}_j^T \, .(\mathbf{z}-\mathbf{c}_j)) \; (\overline{\mathbf{z}-\mathbf{c}_j}) \tag{3.26}$$

where $(\overline{\mathbf{z}-\mathbf{c}_j})$ is the conjugate of the distance between the input $\mathbf{z}$ and center $\mathbf{c}_j$ and $\eta_{\boldsymbol{\sigma}}$ is the learning rate for the weight parameter $\boldsymbol{\sigma}_j$. This learning rate parameter can be real-valued, imaginary-valued or complex-valued, as the parameter $\boldsymbol{\sigma}_j$ is a complex-valued parameter.

Similar derivation for the update of centers of the hidden neurons gives

$$\frac{\partial E}{\partial \mathbf{c}_j} = \nabla \mathbf{c}_j E = \frac{\partial E}{\partial \mathbf{c}_j^R} + i \frac{\partial E}{\partial \mathbf{c}_j^I} \tag{3.27}$$

where the derivative $\frac{\partial E}{\partial \mathbf{c}_j^R}$ is given by

$$\frac{\partial E}{\partial \mathbf{c}_j^R} = \frac{\partial E}{\partial \widehat{y}_k^R} \left[\frac{\partial \widehat{y}_k^R}{\partial a_j} \frac{\partial a_j^R}{\partial p_j} \frac{\partial p_i}{\partial \mathbf{c}_j^R} + \frac{\partial \widehat{y}_k^R}{\partial a_j} \frac{\partial a_j}{\partial q_j} \frac{\partial q_j}{\partial \mathbf{c}_j^R} + \frac{\partial \widehat{y}_k^R}{\partial b_j} \frac{\partial b_j}{\partial p_j} \frac{\partial p_j}{\partial \mathbf{c}_j^R} + \frac{\partial \widehat{y}_k^R}{\partial b_j} \frac{\partial b_j}{\partial q_j} \frac{\partial q_j}{\partial \mathbf{c}_j^R} \right] +$$
$$\frac{\partial E}{\partial \widehat{y}_k^I} \left[\frac{\partial \widehat{y}_k^I}{\partial a_j} \frac{\partial a_j}{\partial p_j} \frac{\partial p_j}{\partial \mathbf{c}_j^R} + \frac{\partial \widehat{y}_k^I}{\partial a_j} \frac{\partial a_j}{\partial q_j} \frac{\partial q_j}{\partial \mathbf{c}_j^R} + \frac{\partial \widehat{y}_k^I}{\partial b_j} \frac{\partial b_j}{\partial p_j} \frac{\partial p_j}{\partial \mathbf{c}_j^R} + \frac{\partial \widehat{y}_k^I}{\partial b_j} \frac{\partial b_j}{\partial q_j} \frac{\partial q_j}{\partial \mathbf{c}_j^R} \right] \tag{3.28}$$

and $\frac{\partial E}{\partial \mathbf{c}_j^I}$ is given by

$$\frac{\partial E}{\partial \mathbf{c}_j^I} = \frac{\partial E}{\partial \hat{y}_k^R}\left[\frac{\partial \hat{y}_k^R}{\partial a_j}\frac{\partial a_j}{\partial p_j}\frac{\partial p_j}{\partial \mathbf{c}_j^I} + \frac{\partial \hat{y}_k^R}{\partial a_j}\frac{\partial a_j}{\partial q_j}\frac{\partial q_j}{\partial \mathbf{c}_j^I} + \frac{\partial \hat{y}_k^R}{\partial b_j}\frac{\partial b_j}{\partial p_j}\frac{\partial p_j}{\partial \mathbf{c}_j^I} + \frac{\partial \hat{y}_k^R}{\partial b_j}\frac{\partial b_j}{\partial q_j}\frac{\partial q_j}{\partial \mathbf{c}_j^I}\right] + \frac{\partial E}{\partial \hat{y}_k^I}\left[\frac{\partial \hat{y}_k^I}{\partial a_j}\frac{\partial a_j}{\partial p_j}\frac{\partial p_j}{\partial \mathbf{c}_j^I} + \frac{\partial \hat{y}_k^I}{\partial a_j}\frac{\partial a_j}{\partial q_j}\frac{\partial q_j}{\partial \mathbf{c}_j^I} + \frac{\partial \hat{y}_k^I}{\partial b_j}\frac{\partial b_j}{\partial p_j}\frac{\partial p_j}{\partial \mathbf{c}_j^I} + \frac{\partial \hat{y}_k^I}{\partial b_j}\frac{\partial b_j}{\partial q_j}\frac{\partial q_j}{\partial \mathbf{c}_j^I}\right] \tag{3.29}$$

Identifying the partial derivatives of eq. (3.28) and (3.29) from eq. eq. (3.13), (3.15) and (3.19), it can be derived that

$$\triangle \mathbf{c}_j = -\eta_{\mathbf{c}}\, \delta_k\, \overline{v}_{kj}\, \overline{f_a}\,'(\boldsymbol{\sigma}_j^T.(\mathbf{z}-\mathbf{c}_j))\, \overline{\boldsymbol{\sigma}}_j \tag{3.30}$$

where $\eta_{\mathbf{c}}$ is the learning rate parameter for the neuron centers, which can be real-valued, imaginary valued or complex-valued.

Thus, the complex-valued gradient update rule for the three parameters, (V, C and $\boldsymbol{\sigma}$) of FC-RBF network discussed in section 3.3 can be summarized as

$$\triangle v_{kj} = \eta_{\mathbf{v}}\, \overline{z}_h^j\, \delta_k \tag{3.31}$$

$$\triangle \boldsymbol{\sigma}_j = \eta_{\boldsymbol{\sigma}}\, \delta_k\, \overline{v}_{kj}\, \overline{f_a}\,'(\boldsymbol{\sigma}_j^T(\mathbf{z}-\mathbf{c}_j))\, (\overline{\mathbf{z}-\mathbf{c}_j}) \tag{3.32}$$

$$\triangle \mathbf{c}_j = -\eta_{\mathbf{c}}\, \delta_k\, \overline{v}_{kj}\, \overline{f_a}\,'(\boldsymbol{\sigma}_j^T(\mathbf{z}-\mathbf{c}_j))\, \overline{\boldsymbol{\sigma}}_j \tag{3.33}$$

As stated earlier, the learning rates $\eta_{\mathbf{v}}$, $\eta_{\mathbf{c}}$ and $\eta_{\boldsymbol{\sigma}}$ can be either real-valued or imaginary valued or complex-valued . However, in this study, we only consider real-valued learning rates. As FC-RBF learning algorithm is a batch learning algorithm, the size of the network, the initial values of the free parameters and the learning rates have to be fixed *a priori*. In our study, initially, the number of hidden neurons were arbitrarily chosen and the initial centers were chosen using a procedure similar to that given in [13]. The search for the size of the network and the initial centers for the best performance using this method is an extensive, laborious procedure. To improve the performance with a lesser search time and effort, the K-means clustering algorithm, which is widely used in the selection of the centers for RBF networks [16] is presented in the next section.

3.4.1 Network Initialization: K-means Clustering Algorithm

The K-means clustering algorithm used in this study can be summarized as:

- Begin with h neurons.
- Choose h random samples from the training set and assign them as the neuron centers.
- For each complex-valued input of the training set, find the Euclidean distance from each of the randomly chosen centers

$$d_j = \sqrt{(\mathbf{z}-\mathbf{c}_j)\overline{(\mathbf{z}-\mathbf{c}_j)}} \tag{3.34}$$

  where $\mathbf{z}$ is the input and $\mathbf{c}_j$ is the center of the jth hidden neuron.

- The training samples are clustered depending on the minimum d_j from the centers.
- Calculate the center of each hidden neuron as the average of the samples belonging to that particular cluster. The width of each cluster is also calculated.
- After clustering, if the distance between two neuron centers is lesser than the width of either of the cluster, then merge them into a single neuron.
- Repeat steps 3 to 6 until there is no change in cluster center.

In this section, a FC-RBF network and its gradient-descent based learning algorithm have been presented in detail. The performance of the FC-RBF algorithm was compared against the CRBF, C-ELM (RBF) and the sequential CMRAN algorithms. Irrespective of the activation functions used, all these algorithms are based on the assumption of non-recurrent, uniformly distributed training samples. However, in most practical problems, this assumption is not normally satisfied. Therefore, proper selection of the training data set is required to obtain a good generalization performance. In the next section, we present one such system capable of sample selection to achieve a better generalization performance. The system selects a sample for learning/deletion based on the knowledge acquired by the network at the time of presentation of the sample. This selective participation of samples during training ensures that the network is not overtrained for the training data set and also reduces the computation time.

3.5 Meta-cognitive Fully Complex-valued Radial Basis Function Network

Recent works on human learning by [17, 18, 19] suggest that meta-cognition which empowers the learner with a self-regulated learning mechanism is the best learning strategy. Meta-cognition provides a means to accurately assess one's current knowledge, identify when new knowledge is needed as well as provide strategies to acquire that new knowledge ([20]). In this section, we introduce one such meta-cognitive learning algorithm for the FC-RBF network. The FC-RBF network with the meta-cognitive learning algorithm is referred to as a "Meta-cognitive Fully Complex-valued Radial Basis Function (Mc-FCRBF)" network.

First, we briefly explain the concept of meta-cognition in the context of human learning proposed by [2]. Nelson and Narens proposed a simple model of meta-cognition as shown in Fig. 3.4(a). This model has two components, namely, a cognitive component and a meta-cognitive component. The cognitive component represents the knowledge and the meta-cognitive component has a dynamic model of the cognitive component (a mental simulation of the cognitive component) ([2]). The information flow from the cognitive component to the meta-cognitive component is considered as a monitory signal, while the information flow in the reverse direction is considered as a control signal. In particular, the information flowing from the meta-cognitive component to the cognitive component (control) either changes the state of the cognitive component or changes the cognitive component itself. As a result, one of the following three actions could occur at the cognitive component:

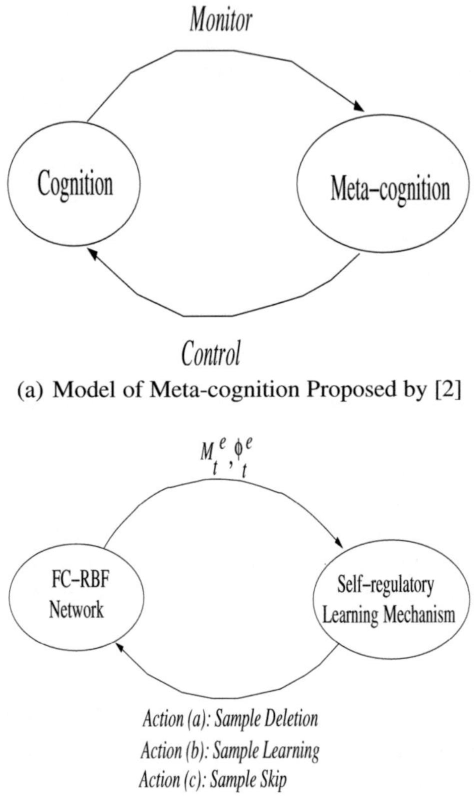

(a) Model of Meta-cognition Proposed by [2]

(b) Meta-cognitive Fully Complex-valued Radial Basis Function network: An Analogy to model of meta-cognition by [2]

Fig. 3.4 Analogy between Model of meta-cognition by [2] and Mc-FCRBF

(a) initiate an action, (b) continue an action, or (c) terminate an action. However, as the control signal does not yield any information from the cognitive component, a monitory signal is needed. The basic notion of monitoring is that the meta-cognitive component is informed about the cognitive component. This changes the state of the meta-cognitive component's model of the cognitive component, including "no change in state". It must be noted that the monitory signal is logically independent of the control signal.

Similar to the [2] model of meta-cognition described above, the meta-cognitive FC-RBF also has two components as shown in the Fig. 3.4(b): FC-RBF network is the cognitive component of Mc-FCRBF and a self-regulatory learning mechanism is its meta-cognitive component. The self-regulatory learning mechanism has a dynamic model of FC-RBF network and *controls* its learning ability by deciding *what-to-learn*, *when-to-learn*, and *how-to-learn*. As a result, when a training sample is presented, one of the following actions occur in FC-RBF network: (a) sample deletion, (b) sample learning, or (c) sample skip. Thus, during the entire training process, the self-regulatory learning mechanism enables selective participation of samples in the training process. Moreover, as each sample is presented, the self-regulatory learning mechanism is informed of the current state of FC-RBF network (*monitor*) through the instantaneous magnitude error and phase error of the sample.

The self-regulatory learning mechanism controls the learning process of FC-RBF network to enable the samples with higher information content to be learnt first and samples with lower information content to be learnt during the later stages of the training process. Samples with similar information content are deleted during the training process. Thus, the meta-cognitive component of Mc-FCRBF prevents learning similar samples in every epoch of the batch learning process, thereby avoiding overtraining and improving the generalization performance of FC-RBF network.

3.5.1 Cognitive Component of Mc-FCRBF: The FC-RBF Network

The cognitive component of Mc-FCRBF is the FC-RBF network with m input neurons, h hidden neurons and n output neurons, as shown in Fig. 3.2. The neurons at the input and output layers are linear, while the neurons at the hidden layer employ the fully complex-valued *sech* activation function. The responses of the neurons in the hidden layer are given by Eq. (3.10). The update rules for the free parameters of the FC-RBF network ($\mathbf{v}_j$, $\boldsymbol{\sigma}_j$ and $\mathbf{c}_j$) are as given in Eqs. (3.31), (3.32) and (3.33).

3.5.2 Meta-cognitive Component of Mc-FCRBF: Self-regulatory Learning Mechanism

In this section, we describe the working principles of the meta-cognitive component of Mc-FCRBF. As shown in Fig. 3.4(b), the meta-cognitive component of Mc-FCRBF controls the learning ability of FC-RBF network (cognitive component) by selecting suitable learning strategies for each sample (control signal) in each epoch of the training process. The instantaneous magnitude and phase errors based on the residual error of FC-RBF network (defined in Eq. (3.14)) for each sample acts as the monitory signal (information flow from cognitive to meta-cognitive component). They are defined by:

- The instantaneous magnitude error:

$$M_t^e = \frac{1}{n}\sqrt{\mathbf{e}^{t^H}.\mathbf{e}^t} \tag{3.35}$$

- The instantaneous phase error:

$$\phi_t^e = \frac{1}{n}\sum_{l=1}^{n}\left|arg(y_l^t) - arg(\widehat{y}_l^t)\right| \tag{3.36}$$

where, the function $arg(.)$ returns the phase of a complex-valued number in $[-\pi, \pi]$, and is given by:

$$arg(z) = atan\left(\frac{imag(z)}{real(z)}\right) \tag{3.37}$$

When a sample is presented to FC-RBF network, the meta-cognitive component decides *what-to-learn*, *when-to-learn*, and *how-to-learn* by taking one of the following three actions (control signals):

Action (a) *Sample Deletion:* Delete those samples from the training data set that contain information similar to that already learnt by the network. This action addresses the *what-to-learn* component of the meta-cognition.

Action (b) *Sample Learning:* Use the sample to update the network parameters in the current epoch. This represents *how-to-learn* the sample in the meta-cognitive framewor .

Action (c) *Sample Skip:* Skip the sample from learning in the current epoch and retain the sample in the training data set, thereby, deciding *when-to-learn* the sample in the context of meta-cognition.

These three actions of the meta-cognitive learning are described in detail below:

- **Sample Deletion:** IF $M_t^e < E_d^M$ AND $\phi_t^e < E_d^\phi$, where E_d^M is the delete magnitude threshold and E_d^ϕ is the delete phase threshold, then the sample t is deleted from the training data set. The thresholds E_d^M and E_d^ϕ are chosen based on the desired accuracy.
- **Sample Learning:** If the sample learning condition given by:

$$\text{IF} \quad M_t^e \geq E_l^M \ \text{OR} \ \phi_t^e \geq E_l^\phi \tag{3.38}$$

is satisfied in the current epoch, then the parameters of the network are updated using the gradient descent based parameter update rules (given in Eq. (3.31), Eq. (3.32), Eq. (3.33)) in the current epoch only. Here, E_l^M is the parameter update magnitude threshold and E_l^ϕ is the parameter update phase threshold. It must be noted that the parameter update magnitude threshold (E_l^M) and parameter update phase threshold (E_l^ϕ) are not fixed. They are self-regulating based on the residual error of the sample in the current epoch, according to the following conditions:

$$\text{IF } M_t^e \geq E_l^M, \text{ THEN } E_l^M := \delta E_l^M - (1-\delta)M_t^e \tag{3.39}$$

$$\text{IF } \phi_t^e \geq E_l^\phi, \text{ THEN } E_l^\phi := \delta E_l^\phi - (1-\delta)\phi_t^e \tag{3.40}$$

where δ is the slope at which the thresholds are self-regulated. Larger value of δ results in a slow decay of the thresholds from their initial values. This helps fewer samples with significant information to be learnt first, and samples containing less significant information to be learnt last. Therefore, larger values of δ ensures that the meta-cognitive principles are emulated efficiently. Usually, δ is set close to 1.

- **Sample Skip:** If a sample does not satisfy the sample deletion or sample learning condition in the current epoch, then the sample is skipped in the current epoch and is retained in the training data set as such. Due to the self-regulating nature of the parameter update thresholds, the sample might be used in learning in subsequent epochs.

The learning algorithm of Mc-FCRBF is summarized in the Pseudocode 1.

Pseudocode 1 Pseudocode: A meta-cognitive Fully Complex-valued Radial Basis Function (Mc-FCRBF) Network.

Input: The data set $\{(\mathbf{z}^1,\mathbf{y}^1),(\mathbf{z}^2,\mathbf{y}^2),\ldots,(\mathbf{z}^N,\mathbf{y}^N)\}$
  of the function to be approximated.
Output: Parameters of the network: $\mathbf{c}_j$, $\mathbf{v}_k$ and $\boldsymbol{\sigma}_j$.
START
  Initialization:
    Choose the number of hidden neurons.
    Initialize $\mathbf{c}_j$, $\mathbf{v}_k$ and $\boldsymbol{\sigma}_j$; $k=1,2,...,h,\ j=1,2,...,n$.
    Initialize the number of epochs OR
    Specify the error based stopping criterion.
  WHILE STOPPING CRITERION(Epoch)
    FOR t = 1,2,..,N(Sample)
      Compute the network output using Eq. (3.11).
      Compute the instantaneous Magnitude error (M_t^e)
      using Eqs.(8.19) and Phase error (ϕ_t^e) using (8.20).
      IF $M_t^e < E_d^M$ AND $\phi_t^e < E_d^\phi$ **THEN**
        Delete the sample from the training data set.
      ELSEIF $M_t^e \geq E_l^M$ OR $\phi_t^e \geq E_l^\phi$ **THEN**
        Update the network parameters
        using Eqs. (3.31), (3.32) and (3.33).
        The thresholds E_l^M and E_l^ϕ are self-regulated
        according to Eqs. (3.39) and (3.40), respectively.
      ELSE
        Reserve the sample in the training data set.
        It may be used in learning/deleted in the
        subsequent epochs.
      ENDIF
    END FOR(Sample)
  END WHILE(Epoch)
STOP

Next, we describe the working principle of Mc-FCRBF using a synthetic complex-valued function approximation problem. The synthetic complex-valued function to be approximated is defined as

$$f_1(\mathbf{z}) = \frac{1}{6}(z_1^2 + z_2^2) \tag{3.41}$$

where z_1 and z_2 are complex-valued variables chosen from within the unit circle and $\mathbf{z}$ is a complex-valued vector ($\mathbf{z} \in \mathbb{C}^2$). For this study, 3000 samples were randomly chosen and used.

We first show the self-regulating nature of the parameter update magnitude (E_l^M) and phase threshold (E_l^ϕ). Fig. 3.5 shows the instantaneous magnitude and phase errors, the parameter update magnitude and phase thresholds, the delete magnitude and phase thresholds over a window of 50 samples (sample instants 1,050-1,100) during epoch 50. Fig. 3.5(a) gives a snapshot of the instantaneous magnitude error, the parameter update and delete magnitude thresholds. The figure clearly shows the self-regulating nature of the parameter update thresholds and the selective participation of samples in the learning process. For example,

- A few samples whose instantaneous magnitude error is greater than the parameter update magnitude threshold (E_l^M) at the time of their presentation to the network are selected for participation in the learning process. (e.g., sample instant 1,067).
- A few samples whose instantaneous magnitude error is less than the delete magnitude threshold (E_d^M) are deleted from the training data set. (e.g., sample instant 1,069).
- There are a few samples whose instantaneous magnitude error is greater than E_d^M and lesser than E_l^M (e.g., sample instant 1,081). These samples neither take part in parameter update, nor are deleted in the current epoch. Instead, they are skipped in the current epoch and retained in the training sample set to be presented to the network in future epochs.

Similarly, Fig. 3.5(b) gives the snapshot of instantaneous phase error, the parameter update phase threshold (E_l^ϕ) and the delete phase threshold (E_d^ϕ) over 50 samples (sample instants 1,050-1,100) during the learning process in epoch 50. The effect of self-regulation based on the instantaneous phase error of the samples is clearly seen from this plot.

- A few samples whose instantaneous magnitude error is greater than the parameter update magnitude threshold (E_l^ϕ) participated in the learning process. (e.g., sample instant 1,065).
- A few samples whose instantaneous phase errors are less than the delete phase threshold (E_d^ϕ) are deleted from the training data set. (e.g., sample instant 1,069).
- There are a few samples whose instantaneous magnitude error is greater than E_d^ϕ and less than E_l^ϕ (e.g., sample instant 1,068). These samples neither took part in learning, nor are deleted in the current epoch. Instead, they are skipped in the current epoch and retained in the training data set, to be presented to the network in the future epochs.

Fig. 3.6 gives the sample history for the number of samples that participated in learning and those that are not used during the learning process over 5,000 epochs. From

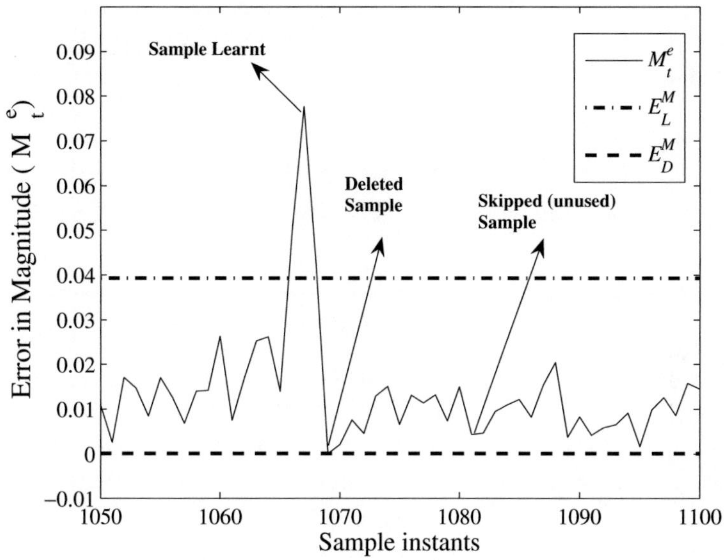

(a) A snap shot of instantaneous magnitude error from 1,050-1,100 sample instants

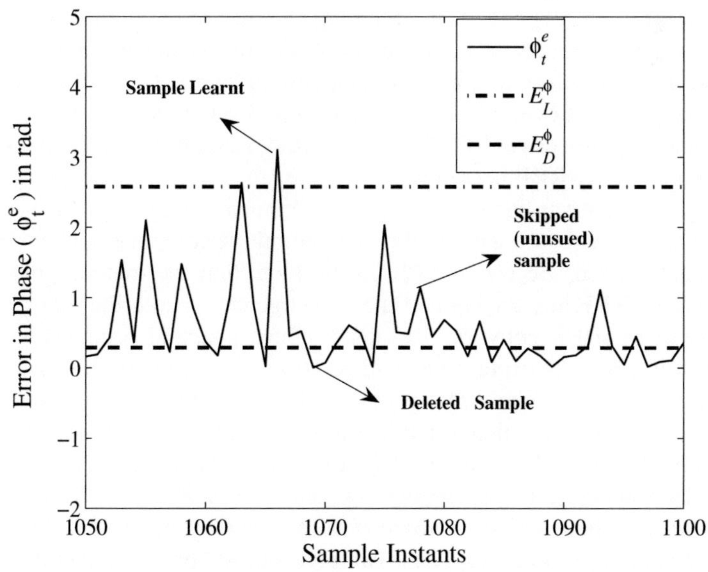

(b) A snap shot of instantaneous phase error from 1,050-1,100 sample instants

Fig. 3.5 Working Principle of the meta-cognitive Component of Mc-FCRBF

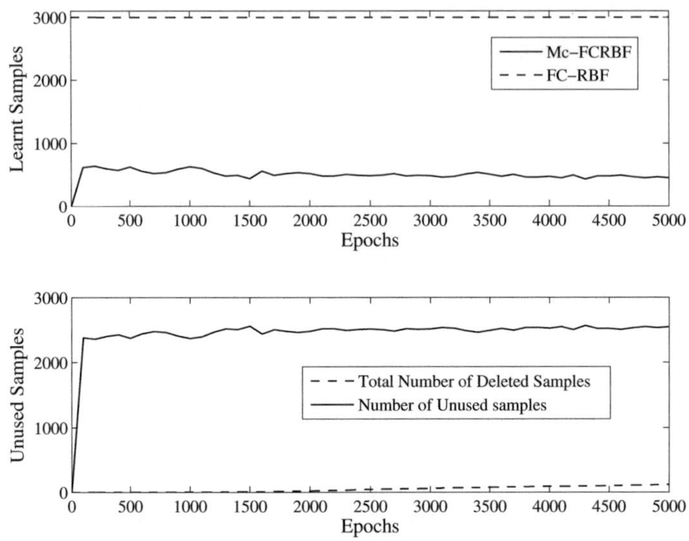

Fig. 3.6 Sample History of Used and Unused Samples

Fig. 3.6, it can be observed that on an average, only 600 samples participated in the learning process in each epoch. The remaining samples are redundant samples that are deleted or samples that do not contribute significantly to the learning process and are reserved in the training sample set without being learnt. Thus, from the discussion in this section, it is evident that the meta-cognitive components of Mc-FCRBF control the learning process of FC-RBF by selecting samples for participation in the learning process of FC-RBF network.

Fig. 3.7 gives the magnitude and phase error convergence plots of FC-RBF network for the CFAP, with and without the meta-cognitive component, over a window of 1,000 epochs. It can be observed from the plots that the meta-cognitive component of Mc-FCRBF has accelerated the convergence of both the magnitude and phase errors of FC-RBF network. The faster convergence also contributes to the overall generalization performance as will be shown later, where the errors of Mc-FCRBF is lower at least by an order compared to that of FC-RBF network. It must be noted here that like any other batch learning algorithm, the performance of Mc-FCRBF is also influenced by the presence of outliers. However, such outliers can be detected by amending the self-regulating conditions suitably. If the current sample error is greater than, say, 6σ (where σ is the standard deviation), then we can classify such samples as outliers. These samples will not participate in the learning process (current epoch). In addition, such samples that consistently produce more than 6σ errors over L consecutive epochs can be deleted from the data set.

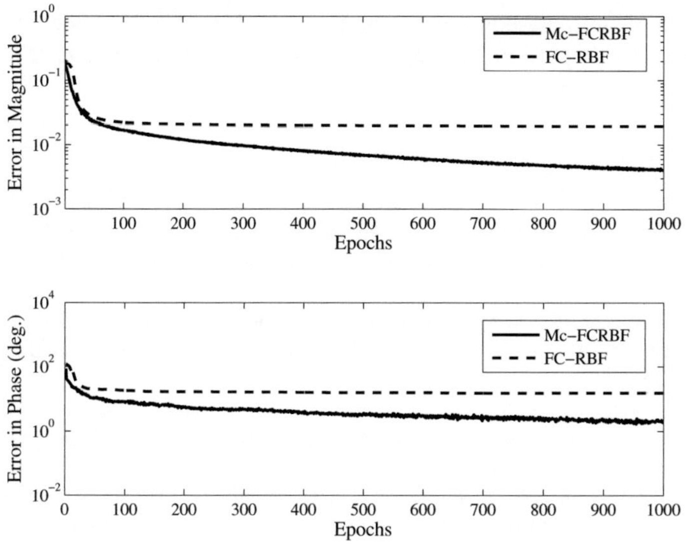

Fig. 3.7 Error convergence plots for CFAP

3.6 Summary

To summarise, in this chapter we have studied complex-valued RBF networks in detail as given below :

- The complex-valued RBF network available in literature uses the real-world Gaussian function that maps $\mathbb{C}^m \rightarrow \mathbb{R}$ as the basis and lacks a fully complex-valued activation function. As the activation function is not fully complex-valued, the gradients used to update the free parameters of the learning algorithm do not consider the complex-valued error/weights. Rather, they consider the real/imaginary components of the error/weights. In doing this, the correlation between the real and imaginary components is lost, and hence, approximation using such a learning algorithm affects the phase approximation of the network significantly.
- Hence, we developed a new fully complex-valued RBF network with a symmetric "*sech*" activation function and derived its gradient descent based learning algorithm. As the activation function and the gradients used in the learning are all fully complex-valued, approximation using FC-RBF network is more accurate.
- A K-Means clustering algorithm chooses the number of neurons, the initial cluster centers and the width of the FC-RBF and this has been presented.
- All the aforementioned algorithms are based on the assumption that the training data is uniformly distributed in the input space. In most practical problems, it is difficult to get uniformly distributed training data with non-recurrent training samples. Hence, a Mc-FCRBF has been developed with the FC-RBF as the

cognitive component and a self-regulatory learning system as the meta-cognitive component. The meta-cognitive component of Mc-FCRBF selects appropriate samples to participate in training in each epoch and also delete samples with similar information.

References

1. Savitha, R., Suresh, S., Sundararajan, N.: A fully complex-valued radial basis function network and its learning algorithm. International Journal of Neural Systems 19(4), 253–267 (2009)
2. Nelson, T.O., Narens, L.: Metamemory: A theoritical framework and new findings. In: Nelson, T.O. (ed.) Metacognition: Core Readings, pp. 9–24. Allyn and Bacon, Boston (1980)
3. Chen, S., McLaughlin, S., Mulgrew, B.: Complex valued radial basis function network,part I: Network architecture and learning algorithms. EURASIP Signal Processing Journal 35(1), 19–31 (1994)
4. Chen, S., McLaughlin, S., Mulgrew, B.: Complex valued radial basis function network, part II: Application to digital communications channel equalization. Signal Processing 36(2), 175–188 (1994)
5. Cha, I., Kassam, S.A.: Channel equalization using adaptive complex radial basis function networks. IEEE Journal on Selected Areas in Communications 13(1), 122–131 (1995)
6. Chen, S., Hong, X., Harris, C.J., Hanzo, L.: Fully complex-valued radial basis function networks: Orthogonal least squares regression and classification. Neurocomputing 71(16-18), 3421–3433 (2008)
7. Li, M.B., Huang, G.-B., Saratchandran, P., Sundararajan, N.: Fully complex extreme learning machine. Neurocomputing 68(1-4), 306–314 (2005)
8. Jianping, D., Sundararajan, N., Saratchandran, P.: Complex-valued minimal resource allocation network for nonlinear signal processing. International Journal of Neural Systems 10(2), 95–106 (2000)
9. Li, M.B., Huang, G.B., Saratchandran, P., Sundararajan, N.: Complex-valued growing and pruning RBF neural networks for communication channel equalisation. IEE Proceedings- Vision, Image and Signal Processing 153(4), 411–418 (2006)
10. Yingwei, L., Sundararajan, N., Saratchandran, P.: A sequential learning scheme for function approximation using minimal radial basis function neural networks. Neural Computation 9(2), 461–478 (1997)
11. Yingwei, L., Sundararajan, N., Saratchandran, P.: Performance evaluation of a sequential minimal radial basis function (RBF) neural network learning algorithm. IEEE Transactions on Neural Networks 9(2), 308–318 (1998)
12. Huang, G.B., Saratchandran, P., Sundararajan, N.: An efficient sequential learning algorithm for growing and pruning RBF (GAP-RBF) networks. IEEE Transactions on Systems, Man, and Cybernetics - Part B: Cybernetics 34(6), 2284–2292 (2004)
13. Suresh, S., Omkar, S.N., Mani, V., Prakash, T.N.G.: Lift coefficient prediction at high angle of attack using recurrent neural network. Aerospace Science and Technology 7(8), 595–602 (2003)
14. Kim, T., Adali, T.: Fully complex multi-layer perceptron network for nonlinear signal processing. Journal of VLSI Signal Processing 32(1/2), 29–43 (2002)
15. Kim, T., Adali, T.: Approximation by fully complex multi-layer perceptrons. Neural Computation 15(7), 1641–1666 (2003)

16. Rollet, R., Benie, G.B., Li, W., Wang, S., Boucher, J.M.: Image classification algorithm based on the RBF neural network and K-means. International Journal of Remote Sensing 19(15), 3003–3009 (1998)
17. Wenden, A.L.: Meta-cognitive knowledge and language learning. Applied Linguistics 19(4), 515–537 (1998)
18. Rivers, W.P.: Autonomy at all costs: An ethnography of meta-cognitive self-assessment and self-management among experienced language learners. The Modern Language Journal 85(2), 279–290 (2001)
19. Isaacson, R., Fujita, F.: Metacognitive knowledge monitoring and self-regulated learning: Academic success and reflections on learning. Journal of the Scholarship of Teaching and Learning 6(1), 39–55 (2006)
20. Josyula, D.P., Vadali, H., Donahue, B.J., Hughes, F.C.: Modeling metacognition for learning in artificial systems. In: Proceedings of 2009 World Congress on Nature and Biologically Inspired Computing (NABIC 2009), pp. 1419–1424 (2009)

Chapter 4
Fully Complex-valued Relaxation Networks

The complex-valued learning algorithms described in the chapters 2 and 3 use a real-valued mean square error function as the performance measure which explicitly minimizes only the magnitude error. In addition, the mean squared error function is non-analytic in the Complex domain (not differentiable in an open set). Therefore, pseudo-gradients, or isomorphic $\mathbb{C}^1 \rightarrow \Re^2$ transformations are generally used in the derivation of the learning algorithms. Use of a mean squared error function which is only an explicit representation of the magnitude error and the the use of pseudo-gradients will affect the phase approximation capabilities of these algorithms. For better phase approximation, one needs to use an error function which simultaneously minimizes both the magnitude and phase errors [1]. But if such functions are inseparable into their real and imaginary parts, then the pseudo gradients are invalid and one needs to identify a mathematical tool to derive the gradients of such functions. Hence, there is a need for the development of a fully complex-valued neural network and its learning algorithm using the fully complex-valued gradients to overcome the above-mentioned issues.

In this chapter, we present a fully complex-valued single hidden layer neural network with a Gaussian like activation function in the hidden layer and an exponential activation function in the output layer. For a given training data set and number of hidden neurons, the network parameters are analytically estimated using a projection based learning algorithm. The learning algorithm employs a nonlinear logarithmic error function as the energy function which explicitly contains both the magnitude and phase errors. The problem of finding the optimal weights is formulated as a nonlinear programming problem. The problem is solved with the help of Wirtinger calculus [2]. Wirtinger calculus provides a framework for determining gradients of a non-analytic nonlinear complex (energy) function. The projection based learning algorithm converts the nonlinear programming problem into solving a system of linear equations and provides a solution for the optimal weights corresponding to the minimum energy point of the energy function. This is similar to the relaxation process, where the system always returns to a minimum energy state from a given initial condition [3]. Therefore, we refer to the proposed complex-valued network as a, 'Fully Complex-valued Relaxation Network (FCRN)'. The projection based

S. Suresh et al.: Supervised Learning with Complex-valued Neural Networks, SCI 421, pp. 73–83.
springerlink.com

learning algorithm of FCRN requires minimal computational effort to approximate any desired function with higher accuracy. In the next sections, we present FCRN and its learning algorithm in detail.

4.1 Fully Complex-valued Relaxation Networks

Complex-valued neural networks are generally employed to handle applications where the signals involved are inherently complex-valued. An efficient complex-valued neural network is required to preserve the nonlinear transformations (both in magnitude and phase) between the complex-valued inputs and their corresponding targets with a minimal computational effort. In this section, we present one such complex-valued neural network and its "projection based learning" algorithm. For a given training data set and the number of hidden neurons, the network parameters are estimated as a solution to a nonlinear programming problem using Wirtinger calculus. The projection based learning algorithm converts the nonlinear programming problem into a system of linear equations and the solution of the same results in computing the optimal output weights. The system of linear equations is derived from a nonlinear logarithmic energy function that contains both the magnitude and phase errors explicitly and the optimal output weights are obtained corresponding to the minimum energy point of this energy function. This process is analogous to a relaxation process, where a system always returns to the minimum energy state from any given initial condition. Hence, the proposed complex-valued neural network is referred to as a, 'Fully Complex-valued Relaxation Network (FCRN)'. The architecture and the learning algorithm of FCRN is described in detail below.

4.1.1 FCRN Architecture

The fully complex-valued relaxation network is a single hidden layer complex-valued neural network having a linear input layer with m neurons, a non-linear hidden layer with h neurons and a non-linear output layer with n neurons, as shown in Fig. 4.1. The neurons in the hidden layer employ a hyperbolic secant function ($sech$) whose magnitude response is similar to that of the real-valued Gaussian activation function.

For a given m-dimensional input $\mathbf{z}^t = [z_1^t, \cdots, z_m^t]^T \in \mathbb{C}^m$, the response of the k-th hidden neuron ($z_h^{k^t}$) is given by:

$$z_h^{k^t} = sech\left(\boldsymbol{\sigma}_k^T\left(\mathbf{z}^t - \mathbf{c}_k\right)\right);\ k = 1, \cdots, h \tag{4.1}$$

where $\boldsymbol{\sigma}_k = [\sigma_k^1, \cdots, \sigma_k^m]^T \in \mathbb{C}^m$ is the scaling factor of the k-th hidden neuron, $\mathbf{c}_k = [c_k^1, \cdots, c_k^m]^T \in \mathbb{C}^m$ is the center of the k-th hidden neuron, the superscript T represents the transpose operator, and $sech(z) = \frac{2}{e^z + e^{-z}}$.

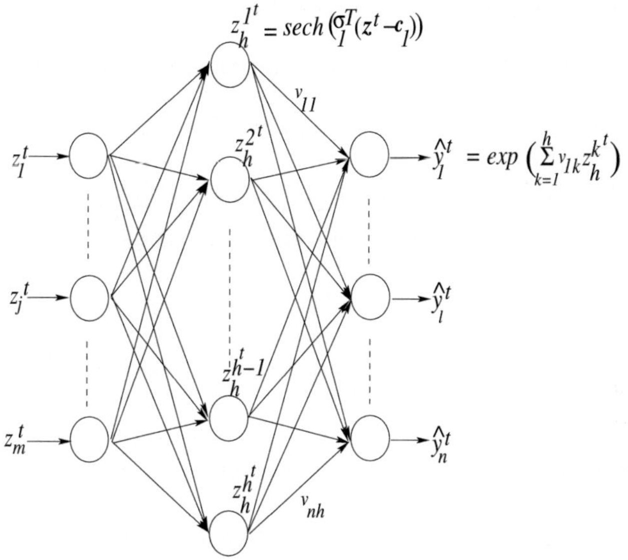

Fig. 4.1 The architecture of FCRN

The n-dimensional output of the network is given by $\widehat{\mathbf{y}}^t = [\widehat{y}_1^t, \cdots, \widehat{y}_n^t]^T \in \mathbb{C}^n$. The neurons in the output layer employ an *exponential* activation function and the response of the l-th output neuron is:

$$\widehat{y}_l^t = exp\left(\sum_{k=1}^{K} v_{lk} z_h^{k^t}\right); \; l = 1, \cdots, n \tag{4.2}$$

where $v_{lk} \in \mathbb{C}$ is the weight connecting the k-th hidden neuron and the l-th output neuron.

Given a training data set $\{(\mathbf{z}^1, \mathbf{y}^1), \cdots, (\mathbf{z}^t, \mathbf{y}^t), \cdots, (\mathbf{z}^N, \mathbf{y}^N)\}$, with $\mathbf{z}^t \in \mathbb{C}^m$ is the m-dimensional input and $\mathbf{y}^t \in \mathbb{C}^n$ is the n-dimensional target of the t-th training sample. The main objective of the fully complex-valued relaxation network is to estimate the free parameters of the network ($\mathbf{C}$, $\boldsymbol{\sigma}$ and $\mathbf{V}$) such that the predicted output ($\widehat{\mathbf{y}}$) is as close as possible to the target output ($\mathbf{y}$), with a given number of hidden neurons (h). In other words, FCRN is required to approximate the underlying transformation function ($\mathbf{F} : \mathbf{z}^t \rightarrow \mathbf{y}^t$) as accurately as possible. Most of the learning algorithms reported in the literature use the mean square error deviation between actual ($\mathbf{y}^t$) and predicted output ($\widehat{\mathbf{y}}^t$) as the performance criterion, which is only an explicit minimization of the magnitude of error. However, accurate estimation of both magnitude and phase of the signals are important in many real-world applications involving complex-valued signals [4]. In this chapter, we use a nonlinear logarithmic error function as the energy function with an explicit representation of both the magnitude and phase of the actual and predicted outputs.

4.1.2 Nonlinear Logarithmic Energy Function

The actual output ($\mathbf{y}^t$) of the t-th training sample is represented in polar form as

$$y_l^t = r_l^t \, exp\left(i\phi_l^t\right); \; l = 1, 2, \cdots, n \tag{4.3}$$

where $r_l^t = ||y_l^t||$ is the magnitude of y_l^t and $\phi_l^t = arctan\left(\frac{Im(y_l^t)}{Re(y_l^t)}\right)$[1] is the phase of y_l^t.[2]

Similarly, the predicted output ($\widehat{\mathbf{y}}^t$) of t-th training sample is represented in polar form as

$$\widehat{y}_l^t = \widehat{r}_l^t \, exp\left(i\widehat{\phi}_l^t\right); \; l = 1, 2, \cdots, n \tag{4.4}$$

where $\widehat{r}_l^t = ||\widehat{y}_l^t||$ is the estimated magnitude and $\widehat{\phi}_l^t = arctan\left(\frac{Im(\widehat{y}_l^t)}{Re(\widehat{y}_l^t)}\right)$ is the estimated phase.

The energy function J_t should be a monotonically decreasing function that represents the magnitude and phase quantities explicitly, i.e., $J_t \to 0$ when $\widehat{\mathbf{r}}^t \to \mathbf{r}^t$ and $\widehat{\boldsymbol{\phi}}^t \to \boldsymbol{\phi}^t$.

We propose an energy function that uses a logarithmic function for explicit representation of both the magnitude and phase of complex signals and is of the form

$$J_t = \sum_{l=1}^{n} \left(ln(\widehat{y}_l^t) - ln(y_l^t)\right) \overline{\left(ln(\widehat{y}_l^t) - ln(y_l^t)\right)} \tag{4.5}$$

where $\overline{(.)}$ is the conjugate of the complex signal (.) and $ln(.)$ represents the natural logarithmic function.

Substituting the polar representation of actual (y_l^t) and predicted output ($\widehat{y}_l^t$), the above equation reduces to

$$J_t = \sum_{l=1}^{n} \left(ln\left(\frac{\widehat{r}_l^t}{r_l^t}\right)^2 + \left(\widehat{\phi}_l^t - \phi_l^t\right)^2 \right) \tag{4.6}$$

It can be observed from Eq. (4.6) that the logarithmic energy function represents the magnitude and phase quantities explicitly and J_t tends to 0, when $\widehat{y}_l^t \to y_l^t$. It must also be noted that the energy function is second order continuously differentiable with respect to the network parameters.

For N training samples, the overall energy is defined as

$$\begin{aligned} J(\mathbf{V}) &= \frac{1}{2} \sum_{t=1}^{N} J_t \\ &= \frac{1}{2} \sum_{t=1}^{N} \sum_{l=1}^{n} \left(ln\left(\frac{\widehat{r}_l^t}{r_l^t}\right)^2 + \left(\widehat{\phi}_l^t - \phi_l^t\right)^2 \right) \end{aligned} \tag{4.7}$$

[1] $y_l^t = Re(y_l^t) + i\, Im(y_l^t)$, where $Re(.)$ and $Im(.)$ refers to the real and imaginary parts of a complex number, respectively

[2] Note that $i = \sqrt{-1}$ is the Complex operator.

In the next section, we derive the projection based learning algorithm of FCRN such that $J(\mathbf{W})$ is minimum.

4.1.3 A Projection Based Learning Algorithm for FCRN

For a given initial condition (i.e., N training samples, h hidden neurons), the projection based learning algorithm finds the network parameters for which the energy function is minimum, i.e., the network achieves the minimum energy point or relaxation point of the energy function.

The hidden neuron centers ($\mathbf{c}_k$) and scaling factors ($\boldsymbol{\sigma}_k$) of FCRN are chosen as random constants and the optimal output weights ($\mathbf{V}^* \in \mathbb{C}^{n\times h}$) are estimated such that the total energy reaches its minimum.

$$\mathbf{V}^* := arg \min_{\mathbf{V}\in\mathbb{C}^{n\times h}} J(\mathbf{V}) \tag{4.8}$$

The problem of estimating the optimal weight is converted to an unconstrained minimization problem ($J(\mathbf{V}) : \mathbb{C}^{n\times h} \to \Re$) involving minimization of the energy function $J(\mathbf{V})$. Let $\mathbf{V}^* \in \mathbb{C}^{n\times h}$, then $\mathbf{V}^*$ is the optimal output weight corresponding to the minimum of the energy function if $J(\mathbf{V}^*) \leq J(\mathbf{V})\ \forall\ \mathbf{V} \in \mathbb{C}^{n\times h}$. The optimal $\mathbf{V}^*$ corresponding to the minimum energy point of the energy function ($J(\mathbf{V}^*)$) is obtained by equating the first order partial derivative of $J(\mathbf{V})$ with respect to the output weight to zero, i.e.,

$$\frac{\partial J(\mathbf{V})}{\partial v_{lp}} = 0;\ \ l = 1,\cdots,n;\ p = 1,\cdots,h \tag{4.9}$$

For convenience, we rewrite the energy function as

$$J(\mathbf{V}) = \frac{1}{2}\sum_{t=1}^{N}\sum_{l=1}^{n}(ln(\widehat{y}_l^t)-ln(y_l^t))\,\overline{(ln(\widehat{y}_l^t)-ln(y_l^t))} \tag{4.10}$$

By substituting the predicted output ($\widehat{\mathbf{y}}^t$) from Eq. (8.17) in Eq. (4.10), the energy function reduces to

$$J(\mathbf{V}) = \frac{1}{2}\sum_{t=1}^{N}\sum_{l=1}^{n}\left(\sum_{k=1}^{K} v_{lk}z_h^{k^t} - ln\ y_l^t\right)\overline{\left(\sum_{k=1}^{K} v_{lk}z_h^{k^t} - ln\ y_l^t\right)} \tag{4.11}$$

where $z_h^{k^t}$ is the response of the k-th hidden neuron for t-th training sample.

Since the energy function is a non-analytic, non-linear real-valued function of the complex-valued output weights and is inseparable into its real and imaginary parts, we use Wirtinger calculus[1] [2] to obtain the first order partial derivatives of the energy function with respect to the complex-valued output weight (v_{lp}). Wirtinger calculus eliminates the stringent conditions of analyticity for the *Complex differentiability* imposed by the Cauchy Riemann conditions. They define the Complex differentiability of almost all functions of interest, including the energy function that maps ($\mathbb{C} \rightarrow \Re$). Although the derivatives defined by Wirtinger calculus do not satisfy the Cauchy Riemann equations, they obey all the rules of calculus (like differentiation of products, chain rule etc.).

Using the Wirtinger calculus and the commutative property of the Complex conjugate operator[2], the first order partial derivative of energy function with respect to w_{lp} ($l = 1,2\cdots,n$ and $p = 1,2,\cdots,h$) is given as:

$$\frac{\partial J(\mathbf{W})}{\partial v_{lp}} = \sum_{t=1}^{N} z_h^{p^t} \left[\sum_{k=1}^{h} \overline{v}_{lk} \overline{z}_h^{k^t} - ln\left(\overline{y}_l^t\right) \right] \tag{4.15}$$

Equating the first partial derivative to zero and re-arranging the Eq. (4.15), we get

$$\sum_{k=1}^{h} \overline{v}_{lk} \sum_{t=1}^{N} z_h^{p^t} \overline{z_h}^{k^t} = \sum_{t=1}^{N} ln\left(\overline{y}_l^t\right) z_h^{p^t} \tag{4.16}$$

Eq. (4.16) can be written as

$$\sum_{k=1}^{h} \overline{v}_{lk} A_{pk} = B_{lp};\ p = 1,\cdots,h;\ l = 1,\cdots,n \tag{4.17}$$

which can be represented in matrix form as

[1] Let $f_R(z_c, \overline{z}_c)$ be a real-valued function of a complex-valued variable $z_c = x_r + iy_r$. Then, the following pair of derivatives are defined by the Wirtinger calculus:

$$\mathbb{R}\text{-derivative of } f_R(z_c, \overline{z}_c) = \frac{\partial f_R}{\partial z_c} |_{\overline{z_c}=\text{constant}} \tag{4.12}$$

$$\overline{\mathbb{R}}\text{-derivative of } f_R(z, \overline{z}_c) = \frac{\partial f_R}{\partial \overline{z_c}} |_{z_c=\text{constant}} \tag{4.13}$$

It is proved in [5] that the $\mathbb{R}-derivative$ (Eq. (4.12)) and the $\overline{\mathbb{R}}-derivative$ (Eq. (4.13)) can be equivalently written as

$$\begin{aligned} \frac{\partial f_R}{\partial z_c} &= \frac{1}{2}\left(\frac{\partial f_R}{\partial x_r} - i\frac{\partial f_R}{\partial y_r}\right) \\ \frac{\partial f_R}{\partial \overline{z}_c} &= \frac{1}{2}\left(\frac{\partial f_R}{\partial x_r} + i\frac{\partial f_R}{\partial y_r}\right) \end{aligned} \tag{4.14}$$

where the partial derivatives with respect to x_r and y_r are *true* partial derivatives of the function $f_R(z_c) = f_R(x_r, y_r)$, which is differentiable with respect to the x_r and y_r.

[2] $\overline{z_a + z_b} = \overline{z_a} + \overline{z_b}$ and $\overline{ln(z_a)} = ln(\overline{z_a})$

$$\overline{\mathbf{V}}\mathbf{A} = \mathbf{B} \tag{4.18}$$

where the projection matrix $\mathbf{A} \in \mathbb{C}^{h\times h}$ is given by

$$A_{pk} = \sum_{t=1}^{N} z_h^{p^t} \overline{z_h}^{k^t};\ p = 1,\cdots,h;\ k = 1,\cdots,h \tag{4.19}$$

and the output matrix $\mathbf{B} \in \mathbb{C}^{n\times h}$ is

$$B_{lp} = \sum_{t=1}^{N} ln\ \bar{y}_l^t z_h^{k^t};\ l = 1,\cdots,n;\ p = 1,\cdots,h \tag{4.20}$$

Eq. (4.17) gives the set of $n\times h$ linear equations with $n\times h$ unknown output weights $\mathbf{V}$. Note that the projection matrix is always a square matrix of order $h\times h$.

We state the following propositions to find the closed-form solution for these set of linear equations.

Proposition 4.1. *The responses of the neurons in the hidden layer are unique. i.e.* $\forall\ \mathbf{z}^t$*, when* $k \neq p$, $z_h^{k^t} \neq z_h^{p^t}$; $k,p = 1,2\cdots h, t = 1,\cdots h$.

Proof. Let us assume that

$$\text{For a given } \mathbf{z}^t, z_h^{p^t} = z_h^{k^t};\ k \neq p \tag{4.21}$$

This assumption is valid if and only if

$$\begin{aligned} sech\left(\boldsymbol{\sigma}_p^T(\mathbf{z}^t - \mathbf{c}_p)\right) &= sech\left(\boldsymbol{\sigma}_k^T(\mathbf{z}^t - \mathbf{c}_k)\right) \\ \text{OR}\ \ \boldsymbol{\sigma}_p^T(\mathbf{z}^t - \mathbf{c}_p) &= \boldsymbol{\sigma}_k^T(\mathbf{z}^t - \mathbf{c}_k) \end{aligned} \tag{4.22}$$

The pair of parameters c_{kj} and c_{pj} (that are elements of the vectors $\mathbf{c}_k$ and $\mathbf{c}_p$, respectively), σ_{kj} and σ_{pj} (that are elements of the vectors $\boldsymbol{\sigma}_k$ and $\boldsymbol{\sigma}_p$, respectively) are uncorrelated random constants chosen from a ball of radius 1, i.e.,

$$||c_{kj}|| < 1; ||\sigma_{kj}|| < 1;\ k = 1,\cdots,h;\ j = 1,\cdots,m \tag{4.23}$$

Therefore, $\mathbf{c}_k \neq \mathbf{c}_p$ and $\boldsymbol{\sigma}_k \neq \boldsymbol{\sigma}_p$ for any $\mathbf{z}^t$ (the t-th random input vector of the training data with N samples). Hence, the response of the k-th and p-th hidden neurons are not equal, i.e., $z_h^{p^t} \neq z_h^{k^t}\ \forall\ \mathbf{z}^t;\ t = 1,\cdots,N$.

Proposition 4.2. *The responses of the neurons in the hidden layer are non-zero. i.e* $\forall\ \mathbf{z}$, $z_h^{k^t} \neq 0$; $k = 1,2\cdots h$.

Proof. Let us assume that the hidden layer response of the k-th hidden neuron is 0, i.e.,

$$z_h^{k^t} = 0 \tag{4.24}$$

This is possible if and only if

$$\begin{aligned} \mathbf{c}_k^T(\mathbf{z}^t - \boldsymbol{\sigma}_k) &= \infty \\ \mathbf{z} \to \infty, \text{ or } &\mathbf{c}_k \to \infty, \text{ or } \boldsymbol{\sigma}_k \to \infty \end{aligned} \tag{4.25}$$

As stated in Eq. (4.23), the hidden layer parameters are random constants chosen from within a circle of radius 1. The input variables $\mathbf{z}^t$ are also normalized in a circle of radius 1 such that

$$|z_j| < 1;\ j = 1, \cdots, m \tag{4.26}$$

Hence, the assumption in Eq. (4.24) is not valid for all $\mathbf{z}^t$. Thus, the responses of the neurons in the hidden layer $z_h^{k^t} \neq 0 \,\forall\, \mathbf{z}^t$.

Using the *Proposition* 4.1 and *Proposition* 4.2, we state the following theorem.

Theorem 4.1. *The projection matrix* **A** *is a positive definite Hermitian matrix, and hence, it is invertible.*

Proof. From the definition of the projection matrix **A** given in Eq. (4.19),

$$\mathbf{A}_{pk} = \sum_{t=1}^{N} z_h^{p^t}\, \overline{z_h}^{k^t};\ p = 1, \cdots, h;\ k = 1, \cdots, h \tag{4.27}$$

it can be derived that the diagonal elements of the **A** for the t-th sample is:

$$A_{kk}^t = z_h^{k^t}\, \overline{z_h}^{k^t};\ k = 1, \cdots, h \tag{4.28}$$

From *Proposition* 4.2, the responses of the hidden neurons are non-zero. Hence, $A_{kk}^t \neq 0$. Therefore Eq. (4.28) can be written as

$$A_{kk}^t = |z_h^{k^t}|^2 > 0 \tag{4.29}$$

Hence the diagonal elements of the projection matrix are real, and positive, i.e., $A_{kk}^t \in \Re > 0$. This can be extended for the entire training sample set as:

$$\mathbf{A}_{kk} = \sum_{t=1}^{N} A_{kk}^t \in \Re > 0 \tag{4.30}$$

The off-diagonal elements of the projection matrix (**A**) for the t-th sample is:

$$A_{kj}^t = z_h^{k^t}\, \overline{z_h}^{j^t} \text{ and } A_{jk}^t = z_h^{j^t}\, \overline{z_h}^{k^t} \tag{4.31}$$

$$\Longrightarrow A_{kj}^t = \overline{A}_{jk}^t \tag{4.32}$$

Using the commutative property of the complex conjugate operator, Eq. (4.31) can be extended for all the N samples as:

$$\mathbf{A}_{kj} = \sum_{t=1}^{N} A_{kj}^{t} = \sum_{t=1}^{N} \overline{A}_{jk} = \overline{\mathbf{A}}_{jk} \tag{4.33}$$

From Eqs. 4.30 and 4.33, it can be inferred that the projection matrix **A** is a Hermitian matrix.

A Hermitian matrix is positive definite *iff* for any $\mathbf{q} \neq 0$, $\mathbf{q}^H \mathbf{A} \mathbf{q} > 0$. Let us consider an unit basis vector $\mathbf{q}_1 \in \Re^{K \times 1}$ such that $q_{11} = 1$ and $q_{12} \cdots q_{1K} = 0$, i.e., $\mathbf{q}_1 = [1 \cdots 0 \cdots 0 \cdots 0]^T$. Therefore,

$$\mathbf{q}_1^H \mathbf{A} \mathbf{q}_1 = \mathbf{A}_{11} \tag{4.34}$$

In Eq. (4.30), it was shown that $k = 1, \cdots, h, \mathbf{A}_{kk} \in \Re > 0$. Therefore,

$$\mathbf{A}_{11} \in \Re > 0 \implies \mathbf{q}_1^H \mathbf{A} \mathbf{q}_1 > 0 \tag{4.35}$$

Similarly, for an unit basis vector $\mathbf{q}_k = [0 \cdots 1 \cdots 0]^T$, the product $\mathbf{q}_k^H \mathbf{A} \mathbf{q}_k$ is given by

$$\mathbf{q}_k^H \mathbf{A} \mathbf{q}_k = \mathbf{A}_{kk} > 0; \; k = 1, \cdots, h \tag{4.36}$$

Let $\mathbf{p} \in \mathbb{C}^h$ be the linear transformed sum of h such unit basis vectors, i.e., $\mathbf{p} = \mathbf{q}_1 t_1 + \cdots + \mathbf{q}_k t_k + \cdots + \mathbf{q}_h t_h$, where $t_k \in \mathbf{C}$ is the transformation constant. Then,

$$\begin{aligned} \mathbf{p}^H \mathbf{A} \mathbf{p} &= \sum_{k=1}^{h} (\mathbf{q}_k t_k)^H \mathbf{A} \sum_{k=1}^{h} (\mathbf{q}_k t_k) \\ &= \sum_{k=1}^{h} |t_k|^2 \left(\mathbf{q}_k^H \mathbf{A} \mathbf{q}_k \right) \\ &= \sum_{k=1}^{h} |t_k|^2 \mathbf{A}_{kk} \end{aligned} \tag{4.37}$$

As shown in Eq. (4.30), $\mathbf{A}_{kk} \in \Re > 0$. Also, that $|t_k|^2 \in \Re > 0$ is evident. Hence,

$$\begin{aligned} & |t_k|^2 \mathbf{A}_{kk} \in \Re > 0 \; \forall \, k \\ \implies & \sum_{k=1}^{h} |t_k|^2 \mathbf{A}_{kk} \in \Re > 0 \end{aligned} \tag{4.38}$$

Thus, the projection matrix A is positive definite, and is hence, invertible.

The solution for **V** obtained as a solution to the set of equations, given in Eq. (4.18) is minimum, if $\frac{\partial^2 J}{\partial {v_{lp}}^2} > 0$. The second derivative of the energy function (J) with respect to the output weights is given by,

$$\frac{\partial^2 J(\mathbf{V})}{\partial {v_{lp}}^2} = \sum_{t=1}^{N} z_h^{p^t} \overline{z_h}^{k^t} = \sum_{t=1}^{N} |z_h^{p^t}|^2 > 0 \tag{4.39}$$

As the second derivative of the energy function $J(\mathbf{V})$ is positive, the following observations can be made from Eq. (4.39):

1. The function J is a convex function.
2. The output weight V^* obtained as a solution to the set of linear equations (Eq. (4.18)) is the weight corresponding to the minimum energy point of the energy function (J).

Using the *Theorem 1*, the solution for the system of equations in Eq. (4.18) can be determined as follows:

$$\overline{\mathbf{V}}^* = \mathbf{B}\mathbf{A}^{-1} \tag{4.40}$$

Applying the commutative law of multiplication of complex-valued conjugates,

$$\mathbf{V}^* = \overline{\mathbf{B}}\,\overline{\mathbf{A}}^{-1} \tag{4.41}$$

Thus the estimated output weights ($\mathbf{V}^*$) corresponds to the minimum energy point of the energy function.

The projection based learning algorithm of FCRN can be summarized as:
Given the training data set: $\left\{(\mathbf{z}^1,\mathbf{y}^1),...(\mathbf{z}^t,\mathbf{y}^t),...,(\mathbf{z}^N,\mathbf{y}^N)\right\}$
Step 1: Choose the number of hidden neurons: h and the random hidden layer parameters: $\mathbf{c}_k, \boldsymbol{\sigma}_k$; $k = 1,\cdots,h$
Step 1: Compute the hidden layer responses h_k^t using

$$z_h^{k^t} = sech\left(\boldsymbol{\sigma}_k^T\left(\mathbf{z}^t - \mathbf{c}_k\right)\right); \;\; k = 1,\cdots,h \tag{4.42}$$

Step 2: Compute the projection matrix $\mathbf{A}$ using

$$\mathbf{A}_{pk} = \sum_{t=1}^{N} z_h^{p^t}\,\overline{z_h}^{k^t}; \;\; p,k = 1,\cdots,h \tag{4.43}$$

Step 2: Compute the output matrix B using

$$\mathbf{B}_{lp} = \sum_{t=1}^{N} ln\left(\bar{y}_l^t\right) z_h^{p^t}; \;\; l = 1,\cdots,n; \;\; p = 1,\cdots,h \tag{4.44}$$

Step 3: Compute the optimum output weights using

$$\mathbf{W}^* = \overline{\mathbf{B}}\,\overline{\mathbf{A}}^{-1} \tag{4.45}$$

4.2 Summary

In this chapter, we have presented a projection based learning algorithm for a fully complex-valued relaxation network. For a given set of hidden layer neuron and their associated parameters, the projection based learning algorithm determines the output weights corresponding to the minimum energy point of an energy function which nonlinear, logarithmic and uses an explicit representation of both the magnitude and phase of the target and the predicted outputs. Using the Wirtinger calculus, the output weights of FCRN are determined as a solution to a nonlinear programming problem. The projection based learning algorithm computes the optimal output weights by converting the nonlinear programming problem to a problem of solving a set of linear equations. The output weights thus obtained are optimum and the training time required for learning is minimum.

In the next chapter, we evaluate the performances of the algorithms discussed in Chapters 2-4: FC-MLP, IC-MLP, FC-RBF, Mc-FCRBF and FCRN on a set of complex-valued function approximation problems.

References

1. Savitha, R., Suresh, S., Sundararajan, N., Saratchandran, P.: A new learning algorithm with logarithmic performance index for complex-valued neural networks. Neurocomputing 72(16-18), 3771–3781 (2009)
2. Wirtinger, W.: Zur formalen theorie der funktionen von mehr komplexen vernderlichen. Annals of Mathematics 97 (1927)
3. Yu, S.-S., Tsai, W.-H.: Relaxation by the Hopfield neural network. Pattern Recognition 25(2), 197–209 (1992)
4. Loss, D.V., de Castro, M.C.F., Franco, P.R.G., de Castro, E.C.C.: Phase transmittance RBF neural networks. Electronics Letters 43(16), 882–884 (2007)
5. Remmert, R.: Theory of Complex Functions. Springer, New York (1991)

Chapter 5
Performance Study on Complex-valued Function Approximation Problems

In this chapter, we evaluate the approximation performances of the fully complex-valued multi-layer perceptron network and the improved fully complex-valued multi-layer perceptron network described in Chapter 2, the fully complex-valued radial basis function network and the meta-cognitive fully complex-valued radial basis function network described in Chapter 3, and the fast learning fully complex-valued relaxation network described in Chapter 4. The performances of these networks are studied in comparison with existing complex-valued learning algorithms like the complex-valued extreme learning machine and the complex-valued minimal resource allocation network using two synthetic, complex-valued function approximation problems and two real-world problems. The real world problems consist of a Quadrature Amplitude Modulation (QAM) channel equalization problem with circular signals and an adaptive beam-forming problem with non-circular signals.

5.1 Synthetic Function Approximation Problems

In this section, we define two synthetic complex-valued function approximation problems which were used to study different activation functions and also the performances of the algorithms presented in chapters xxxx , viz., IC-MLP, FC-RBF and Mc-FCRBF in comparison with other existing complex-valued learning algorithms. Hereafter, the two complex function approximation problems are referred to as CFAP-I and CFAP-II,. In these problems, all the function variables are chosen from within a circle of radius of 2.5. The following performance metrics are used to compare the performances of the different complex-valued learning algorithms:

To evaluate the performances of different complex-valued networks, we define the root mean square magnitude error (J_{Me}) and the average absolute phase error (ϕ_e) as

S. Suresh et al.: Supervised Learning with Complex-valued Neural Networks, SCI 421, pp. 85–107.
springerlink.com

$$J_{Me} = \sqrt{\frac{1}{N}\sum_{t=1}^{N}\left[\frac{1}{n}\sum_{k=1}^{n}(e_k^t.\bar{e}_k^t)\right]} \tag{5.1}$$

$$\phi_e = \frac{1}{N}\sum_{t=1}^{N}\left[\frac{1}{n}\left[\sum_{k=1}^{n}\mid[arg(y_k^t) - arg(\hat{y}_k^t)]\mid\right]\right] \text{ in deg.} \tag{5.2}$$

where, N is number of samples in the training set. For each initializing value, the root mean square magnitude error J_{Me} and average absolute phase error ϕ_e are computed along with a statistical analysis on different runs. It should be noted that in eq. (5.2), the arg operator returns the angle of the complex-valued signal as: $arg(y_k^t) = tan^{-1}\frac{imag(y_k^t)}{real(y_k^t)}$ in $rad.$, where $real(y_k^t)$ and $imag(y_k^t)$ are the Real and Imaginary components of y_k^t respectively.

Since the algorithms described in Chapters 2, 3 and 4 are batch learning algorithms, the following procedure is used to decide the number of neurons at the hidden layer:

Procedure for selection of network configuration

Step 1. Select a network with a minimum configuration ($h = m + n$).
Step 2. Initialize the network parameters with uniform distribution (zero mean and low variance) of the parameters in the all the four quadrants.
Step 3. Train the network until the training magnitude error (J_{Me}) < 0.002 and phase error (ϕ_e) $< 0.3^o$ or until 5000 epochs.
Step 4. Calculate the magnitude (J_{Me}) and phase (ϕ_e) error for the testing samples.
Step 5. If $J_{Me} > 0.002$ and $\phi_e > 0.3^o$,

- If $h < h_{max}$, then increase the number of hidden neurons by 2 and return to step 2.

Else, stop.

5.1.1 Synthetic Complex-valued Function Approximation Problem I (CFAP-I)

The synthetic complex-valued function to be approximated is defined as

$$f_1(\mathbf{z}) = \frac{1}{6}(z_1^2 + z_2^2) \tag{5.3}$$

where z_1 and z_2 are complex-valued variables chosen from within the unit circle and $\mathbf{z}$ is a complex-valued vector ($\mathbf{z} \in \mathbb{C}^2$). For performance study using the CFAP-I, 3000 samples for training and 1000 samples for testing are randomly generated and used.

Table 5.1 presents the results of the performance study on CFAP-I. From the results, it can be observed that the proposed algorithms, viz., IC-MLP, FC-RBF, and Mc-FCRBF outperforms the existing complex-valued learning algorithms in the literature, viz., CRBF, CMRAN, C-ELM and FC-MLP. Based on the results, the following observations emerge:

- Although the magnitude approximation ability of IC-MLP is similar to that of FC-MLP, its phase approximation ability has improved drastically (testing phase error of IC-MLP is at least 40% lower than that of FC-MLP).
- The magnitude and phase approximation performances of FC-RBF is better than other complex-valued radial basis function networks available in the literature, viz., CRBF, CMRAN and C-ELM with a RBF activation function. Further, FC-RBF requires only a fewer neurons to achieve this higher performance.
- FC-RBF with KMC selects the right number of neurons and initialization to achieve an approximation performance better than that of FC-RBF. The network size is also smaller than that of FC-RBF.
- The meta-cognitive component of Mc-FCRBF boosts the magnitude and phase approximation performance of FC-RBF. Moreover, it uses only 520 samples per epoch on an average to achieve this performance. This also results in considerable reduction of training time.

Table 5.1 CFAP-I: Performance comparison of various complex-valued learning algorithms

Network	h^*	Training Error		Testing Error	
		J_{Me}	ϕ_e (deg.)	J_{Me}	ϕ_e (deg.)
CRBF	20	0.592	45.32	0.623	47.15
CMRAN	27	0.0594	13.88	0.0614	18.83
C-ELM (RBF)	20	0.6886	34.89	0.704	36.15
FC-MLP	15	0.002	0.29	0.003	0.163
IC-MLP$^@$	15	0.002	0.05	0.003	0.09
FC-RBF	15	0.0019	0.338	0.003	0.3438
FC-RBF with KMC	11	0.00142	0.5563	0.00137	0.1941
Mc-FCRBF$^{@@}$	15	0.005	0.3	0.0016	0.18

* Number of hidden neurons $^@$ $k_1 = 25$; $k_2 = 2$
$^{@@}$ On an average, 520 samples participated in learning in all the epochs.

From the results in Table 5.1, it may be noted that the fully complex-valued neural networks with fully complex-valued activation functions outperform existing algorithms. Next, we evaluate the performance of FCRN and other algorithms using another synthetic complex-valued function approximation problem.

5.1.2 Synthetic Complex-valued Function Approximation Problem II (CFAP-II)

For this problem, the function to be approximated is given by eq. (5.4).

$$f_2(\mathbf{z}) = \frac{1}{1.5}(z_3 + 10z_1z_4 + \frac{z_2^2}{z_1}) \tag{5.4}$$

where z_1, z_2, z_3 and z_4 are complex-valued variables chosen from within the unit circle and $\mathbf{z}$ is a complex-valued vector ($\mathbf{z} \in \mathbb{C}^4$). A training sample set with 3000 samples and testing sample set with 1000 samples were randomly generated for this study.

Table 5.2 presents the performance results of the complex-valued learning algorithms considered in this study on the CFAP-II. From the results, it can be observed that the proposed algorithms, viz., IC-MLP, FC-RBF, and Mc-FCRBF outperform existing complex-valued learning algorithms in the literature, viz., CRBF, CMRAN, C-ELM and FC-MLP. The following observations emerge:

- The magnitude and phase approximation abilities of IC-MLP are better than that of FC-MLP (testing phase error of IC-MLP is at least 80% lower than that of FC-MLP).
- The magnitude and phase approximation performance of FC-RBF is better than other complex-valued radial basis function networks available in the literature, viz., CRBF, CMRAN and C-ELM with a RBF activation function.
- FC-RBF with KMC chooses the lowest number of neurons and best initialization to achieve an approximation performance better than that of FC-RBF. The network size is also smaller than that of FC-RBF.
- The meta-cognitive component of Mc-FCRBF boosts the magnitude and phase approximation performance of FC-RBF. Moreover, on an average, it uses only

Table 5.2 CFAP-II: Performance comparison of various complex-valued learning algorithms

Algorithm	h^*	Training Error		Testing Error	
		J_{Me}	ϕ_e (deg.)	J_{Me}	ϕ_e (deg.)
CRBF	15	0.1465	51.18	0.18155	51.96
C-ELM (RBF)	15	0.1917	90.06	0.22974	88.17
CMRAN	14	0.0257	2.23	0.476	18.68
FC-MLP	15	0.027	15.74	0.0544	15.6
IC-MLP	15	0.02	0.46	0.04	1.14
FC-RBF	20	0.0196	15.87	0.0478	15.83
FC-RBF with KMC	11	0.0114	3.12	0.0909	13.59
Mc-FCRBF	15	0.0009	0.53	0.0009	0.56
FCRN	10	0.03	1.38	0.06	3.22

* Number of hidden neurons

600 samples in each epoch of the training process to achieve a much better performance. This also results in a considerable reduction of the training time.
- The phase approximation performance of FCRN is much better than that of FC-RBF without the meta-cognitive component, although the magnitude approximation performance is comparable to that of FC-RBF.

5.2 Real-World Problems

5.2.1 *Complex Quadrature Amplitude Modulation Channel Equalization Problem*

Quadrature amplitude modulation is both an analog and a digital modulation scheme. It conveys two analog message signals, or two digital bit streams, by changing (modulating) the amplitudes of two carrier waves, using the amplitude-shift keying digital modulation scheme or amplitude modulation analog modulation scheme. These two waves, usually sinusoids, are out of phase with each other by 90° and are thus called quadrature carriers or quadrature components. The modulated waves are summed, and the resulting waveform is a combination of both phase-shift keying and amplitude-shift keying, or in the analog case of phase modulation and amplitude modulation. In the digital QAM case, a finite number of at least two phases and at least two amplitudes are used. Thus, the signals in the QAM schemas are complex-valued. When the QAM signals are transmitted over a channel, the nonlinear characteristics of the channel cause spectral spreading, inter-symbol interference and constellation warping. Hence, an equalizer is essential at the receiver of the communication channel to reduce the precursor inter-symbol interference without any substantial degradation in the signal-to-noise ratio.

As the signals are all complex-valued, of equal magnitude, differing only by the phase, accurate phase approximation of the signals becomes essential. Hence, the equalization ability of the different complex-valued neural networks and their learning algorithm are studied in detail. The efficiency of each equalizer to isolate noisy signals from the data symbol is measured in terms of its Symbol Error Rate (SER) that is defined as the percentage of symbols that have errors relative to the total number of symbols received in a transmission system, usually expressed as ten to a negative power.

$$SER = \left(\frac{\text{Number of errors in predicted symbols}}{\text{Total number of symbols}} \right) \tag{5.5}$$

Problem formulation: Consider the base band discrete time model of a data transmission system [1] given by

$$y_q(n) = \widehat{y}_q(n) + e_q(n) = f_h(s(n), s(n-1), \cdots s(n-n_h+1)) + e_q(n) \tag{5.6}$$

where a complex-valued digital sequence $s(n)$ is transmitted through a dispersive Complex channel, the channel output $y_q(n)$ is corrupted by an additive complex-valued noise $e_q(n)$ and $\widehat{y}_q(n)$ is the function (which may be linear or nonlinear) of the digital sequence and the channel given by $f_h(s(n), s(n-1), \cdots s(n-n_h+1))$, n_h is the order of the finite impulse response channel. The task of the symbol decision equalizer is to reconstruct the transmitted symbols $s(n-\tau)$ based on noisy channel observation vector $\mathbf{y}_q(n) = [y_q(n) \cdots y_q(n-m+1)]^T$, where m is the equalizer delay (the number of past observations required for reconstruction), or the input dimension of the equalizer.

Assume a 4-QAM input sequence with alphabet $\alpha = \{\alpha^{\{1\}}\}, \alpha^{\{2\}}, \alpha^{\{3\}}, \alpha^{\{4\}}\}$ where $\alpha^{\{1\}} = a + ja$, $\alpha^{\{2\}} = a - ja$, $\alpha^{\{3\}} = -a + ja$, $\alpha^{\{4\}} = -a - ja$ going through a noiseless channel of order n_h. The input sequence: $\mathbf{s}(n) = [s(n) \cdots s(n-m+2-n_h)]^T$ would result in n_s points or values of noise-free channel output vector $\widehat{\mathbf{y}}_q(n) = [\widehat{y}_q(n) \cdots \widehat{y}_q(n-m+1)]^T$ where $n_s = 4^{n_h+m-1}$. These output vectors are referred to as the desired channel states and are partitioned into four different classes $Y^{(j)}_{q,\{m,\tau\}}$, $(1 \le j \le 4)$, according to the value of $s(n-\tau)$. The number of states in $Y^{(1)}_{q,\{m,\tau\}}, Y^{(2)}_{q,\{m,\tau\}}, Y^{(3)}_{q,\{m,\tau\}}, Y^{(4)}_{q,\{m,\tau\}}$ are denoted as n^1_s , n^2_s , n^3_s and n^4_s respectively. Due to the additive white Gaussian noise, the channel outputs will form clusters around each of these desired channel states. Therefore, the noisy observation vector $\mathbf{y}_q(n) = [y_q(n) \cdots y_q(n-m+1)]^T$ is a random process with a conditional Gaussian density function centered at each of the desired channel states. The noisy observation vector is used as the input to the equalizer to determine the transmitted symbol $s(n-\tau)$. The objective of the equalizer in QAM problem is to retrieve the symbols at the receiver from the past observations ($\mathbf{y}_q$). Since here, magnitudes are equal and the symbols differ only in the phase, phase approximation is more important than magnitude approximation. As the complex-valued neural networks are capable of better phase approximations, they are a good choice to solve the problem of QAM equalization. The schematic diagram of a complex-valued neural network based equalizer is presented in Fig. 5.1. Assume that the real and imaginary part of transmitted symbol $s(k)$ is equiprobable and independent sequences. Let λ be the apriori probability of $\mathbf{y}^{\{j\}}_{qi}$, where $\mathbf{y}^{\{j\}}_{qi} \in Y^{(j)}_{q,\{m,\tau\}}$, $(1 \le j \le 4)$. The conditional probability density function ($f^{\{j\}}_b$) of $\mathbf{y}_q(n)$ given $s(n-t) = \alpha^{\{j\}}$ takes the following form:

$$f^{\{j\}}_b = \lambda \sum_{i=1}^{n^j_s} exp(-(\mathbf{y}_q - \mathbf{y}^{\{j\}}_{qi})^H (\mathbf{y}_q - \mathbf{y}^{\{j\}}_{qi})/2\sigma^2_e) \quad 1 \le j \le 4 \tag{5.7}$$

where H (Hermitian) denotes the complex conjugate transposition. The optimal Bayesian equalizer solution is defined as

$$\widehat{s}(n-\tau) = \alpha^{\{j\}} \text{ where } j = argmax_j \left\{ f^{\{j\}}_b \right\} \tag{5.8}$$

This equation also defines the optimum decision boundaries for the partition of equalizer inputs sets. It is clear that these optimum decision boundaries are

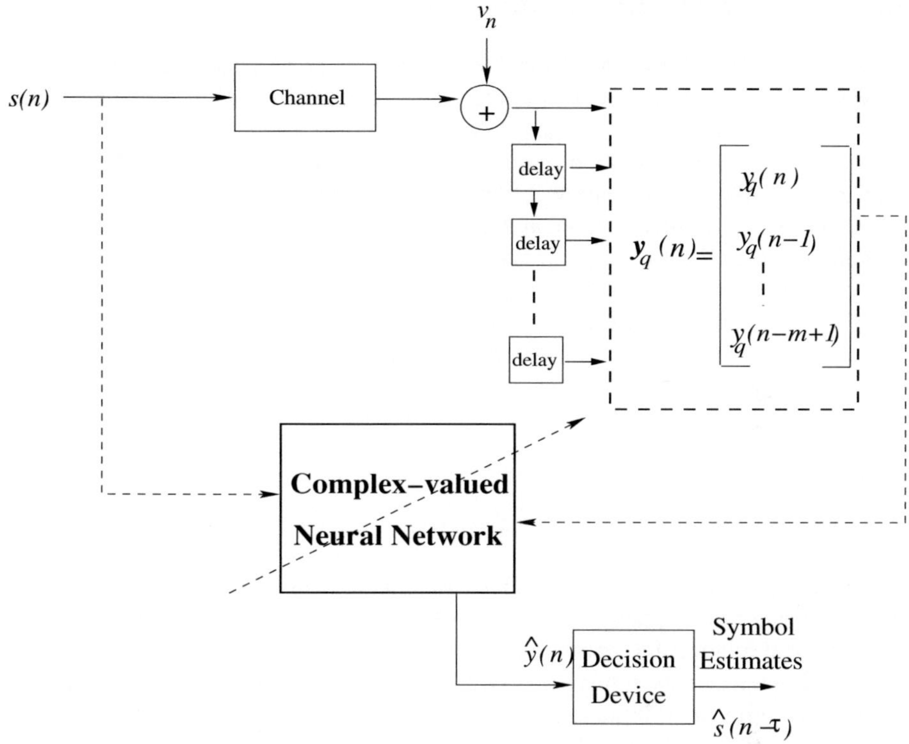

Fig. 5.1 Complex-valued Neural Network Equalization Scheme

nonlinear hyper surfaces in the Complex observation space and realizing such a non-linear boundary will require the equalizer to have non-linear mapping capabilities. Since neural networks are well-known for their non-linear mapping capabilities, they are a good choice for equalizer. Also, since a good equalizer requires accurate phase approximations to classify the signal points in their respective quadrants efficiently, complex-valued neural networks, which are capable of better phase approximations are an excellent choice for non-minimum phase equalization. There are several channel models available to transmit the QAM signals like the real-valued Patra's model [2], complex-valued linear Chen's model [3], the complex-valued non-linear Cha and Kassam channel model [4] etc. The complex-valued non-linear Cha and Kassam channel model is a well-known non-linear channel model for transmission of 4-QAM signals, and to study the non-minimum phase equalizing ability of different equalizers.

In this chapter, we evaluate the performances of various complex-valued neural networks in reconstructing the transmitted symbols from the noisy channel observations, for the Cha and Kassam model [4]. In the next section, we present this model in detail.

5.2.2 Cha and Kassam Channel Model

A well known Complex QAM channel model, the Cha and Kassam model [4] is used to evaluate the performances of various equalizers. Here, the complex-valued neural network based equalizers are used to retrieve one of the 4-QAM signals $\{-1-i,\ -1+i,\ 1-i,\ 1+i\}$ transmitted through the Cha and Kassam channel model. The equalizer uses the channel output at three consecutive instants $(n-2)$, $(n-1)$ and n to predict the transmitted symbol at the time instant $n-1$. Thus, the equalizer uses the channel outputs z_{n-2}, z_{n-1} and z_n to predict the transmitted symbol s_{n-1}. Hence, it is understandable that the equalizer delay is set at $\tau = 1$ as done in [5] and the order of the equalization model is chosen as 3. The output of the Cha and Kassam channel z_n (which is also the input to the equalizer) is given by:

$$z_n = o_n + 0.1 o_n^2 + 0.05 o_n^3 + v_n, \tag{5.9}$$

where

$$o_n = (0.34 - 0.27i) s_n + (0.87 + 0.43i) s_{n-1} + (0.34 - 0.21i) s_{n-2} \tag{5.10}$$

The model is assumed to be affected by a white Gaussian noise (v_n) with zero mean and a variance of 0.01, hence v_n in eq. (5.9) is given by $\aleph(0, 0.01)$. As the inverse mathematical relationship between z_n and $s_{n-\tau}$ does not exist, neural networks can be trained to retrieve the transmitted symbols from the channel output. Thus, a neural network with three input neurons and one output neuron is used in the QAM equalization of the Cha and Kassam model considered. The inputs to the neural network are z_{n-2}, z_{n-1} and z_n, and the targets are s_{n-1}.

In the next section, we present the equalization ability of different complex-valued neural network algorithms presented in chapters 3, 4 and 5 on the channel model presented by Cha and Kassam [4]. Their efficiencies to reject noisy data and to avoid inter-symbol interferences are measured in terms of their symbol error rates. The SER of different complex-valued neural network algorithm based equalizers are plotted at different noise levels, and compared for best equalization performance.

In this section, first we present how the training and testing dataset were generated. Then the various CVNN equalizers are evaluated on the basis of the equalizer framework, i.e., CMLP equalizers, CRBF equalizers and the complex-valued sequential algorithm based equalizers. Finally, the best performance of the best algorithms in the three frameworks are compared and discussed.

Dataset: The different complex-valued neural network equalizers are trained with a pilot sequence of of 5000 randomly generated samples at 20dB SNR. The equalization performances of these equalizers are then tested on a testing sample set of 100000 randomly generated samples for each of the AWGN noise levels ranging between 4dB and 16dB. The generalization ability and the SER performances of the different CVNN equalizers are studied and presented.

Network structure: As mentioned in section 5.2.2, the equalizer input dimension is chosen to be 3. Hence the complex-valued neural networks considered, have 3 input neurons, and as the network is supposed to estimate the input symbol $s_{n-\tau}$, $\tau = 1$, the network was assigned one output neuron. For all the batch learning algorithms considered, viz., SC-MLP, FC-MLP, CRBF and FC-RBF, 15 neurons were chosen for the hidden layer. This was chosen and fixed based on the heuristic procedure discussed in section 5.1 that is similar to the one presented in [6].

Performance measure: The equalization performance of each of the complex-valued neural network algorithm is measured in terms of its symbol error rate performance on the test dataset, i.e., its ability to isolate noisy symbol and inter-symbol interference at different noise levels. Also, the various complex-valued neural network algorithms were evaluated based on the training and testing RMS magnitude error (eq. (5.1)) and average phase errors (eq. (5.2)) in approximating the QAM equalization problem on the Cha and Kassam model [4].

To begin with, the complex-valued equalizers are divided into three categories and their performances studied separately. First, the performances of the various complex-valued MLP equalizers, viz., SC-MLP, FC-MLP and IC-MLP equalizers are presented, followed by the various complex-valued RBF equalizers, viz., CRBF, C-ELM (Gaussian), FC-RBF and Mc-FCRBF equalizers.

5.2.2.1 Complex-valued MLP Equalizers

First, we study the performance of the different complex-valued MLP based equalizers in the batch learning mode. In Table 5.3, the performances of different CMLP based equalizers, viz., SC-MLP [7], FC-MLP [8] with "*asinh*" activation function and IC-MLP (chapter 3) with "*exp*" activation function [9] are tabulated and compared with respect to training and testing magnitude and phase errors, training time and number of neurons required for training. The number of neurons were chosen based on a heuristic procedure discussed in section 5.1, similar to the one presented in [6] for real-valued networks. It is observed that for the QAM problem, the best performances of the network for the three CMLP based algorithms are observed when 15 hidden neurons were used for training. From the table, it can be observed that though the training magnitude and phase error is lesser for FC-MLP equalizer, the testing magnitude and phase errors are high. The testing magnitude error is at least 60% lesser than that of FC-MLP equalizer and the testing phase error is nearly one-third that of FC-MLP. Also, of the three algorithms considered for the complex-valued MLP network, IC-MLP required lesser time for training (atleast 100 seconds lesser than FC-MLP equalizer). Hence, it can be inferred that IC-MLP has better generalization performance than the other MLP learning algorithms considered for this QAM channel equalization problem. The SER plot of the various complex-valued MLP equalizers are presented in Figure 5.2. It can be observed from the figure that the symbol classification efficiency of IC-MLP equalizer is better than that of FC-MLP equalizer, at least by 25% at 16dB SNR. Hence, it can be inferred

Table 5.3 Performance comparison of CMLP equalizers for QAM equalization problem

Algorithm	N_S^*	Time (sec)	h	Training Error		Testing Error	
				J_{Me}	ϕ_e (deg.)	J_{Me}	ϕ_e (deg.)
SC-MLP	5000	4731.25	15	0.377	23.84	0.5502	31.56
FC-MLP	5000	3862.5	15	0.2153	6.47	0.7198	31.1
IC-MLP [1]	5000	3734.4	15	0.228	11.37	0.2278	11.32

* Number of samples used in training
[1] With "*exp*" activation function in the hidden layer.

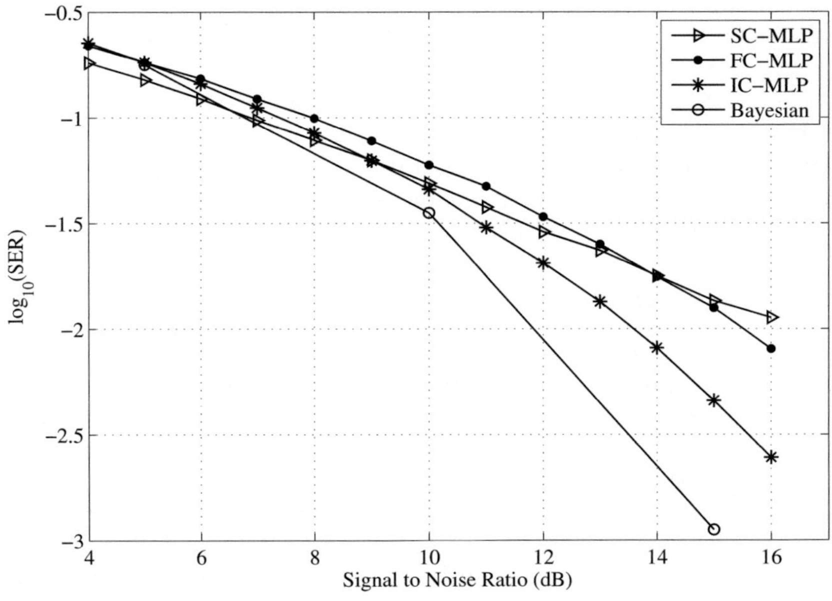

Fig. 5.2 Error Probability Curve for the Complex-valued MLP Based Equalizers

that IC-MLP algorithm proposed in chapter 2 performs better equalization of the channel model considered, than the other CMLP equalizers considered.

5.2.2.2 Complex-valued RBF Equalizers

Next, the performance of the different complex-valued RBF equalizers, viz., CRBF [10] with Gaussian activation function, C-ELM[5] with Gaussian activation function and FC-RBF network [11] with "*sech*" activation function, in batch learning mode, are evaluated for the non-minimal phase equalization of the Cha and Kassam model [4].

In Table 5.4, the QAM equalizer approximation performance of the different complex-valued RBF learning algorithms are evaluated with respect to their training and testing magnitude and phase errors, number of neurons required and the time taken for approximation.

The number of neurons required for training CRBF network and the C-ELM equalizers were chosen based on the heuristic procedure presented in Section 5.1 ([6]). For FC-RBF network, the hidden neurons and the initial centers and weights were chosen using the K-means clustering algorithm presented in section 3.4.1. With proper initialization using the K-means clustering algorithm, the network required only 14 neurons to approximate the QAM equalizer for the Cha and Kassam channel model.

It can be observed from the Table 5.4 that the training and testing magnitude and phase error for FC-RBF network is the least of the three complex-valued RBF algorithms considered. While the testing magnitude error of FC-RBF algorithm is atleast 40% lesser than that of CRBF and C-ELM algorithms, the phase error is nearly about one-third that of the other complex-valued RBF leaning algorithms.

Considering the effect of the self-regulatory system on FC-RBF network, it can be seen that the improvement in phase generalization performance is about 16% and the magnitude approximation performance has improved by nearly 50%. Besides these, the computational time has reduced by nearly one-fourth. This is because, the self-regulatory system chooses only those samples with significant information for learning in each epoch.

Table 5.4 Performance comparison of CRBF equalizers for QAM equalization problem

Algorithm	Time	h	Training Error		Testing Error	
	(sec)		J_{Me}	ϕ_e $(deg.)$	J_{Me}	ϕ_e $(deg.)$
CRBF	8106.6	15	0.5630	35.1911	0.5972	39.86
C-ELM	0.3605	15	0.572	34.14	0.5772	35.11
FC-RBF	3840	15	0.4	31.17	0.41	14.69
Mc-FCRBF	1281.4	15	0.1428	9.57	0.1664	9.539
FCRN	0.998	14	0.3	22.1	0.35	12.62

* Number of samples used in training

In Fig. 5.3, the SER plot for all the complex-valued RBF equalizers are presented, for noise levels ranging from 4dB to 16dB, in order of 1dB increase. It can be observed from this figure also that FC-RBF network presented in chapter 4 classified the QAM signal points efficiently (atleast an advantage of 20% in the SER at 16dB noise level) than the other complex-valued RBF algorithms considered.

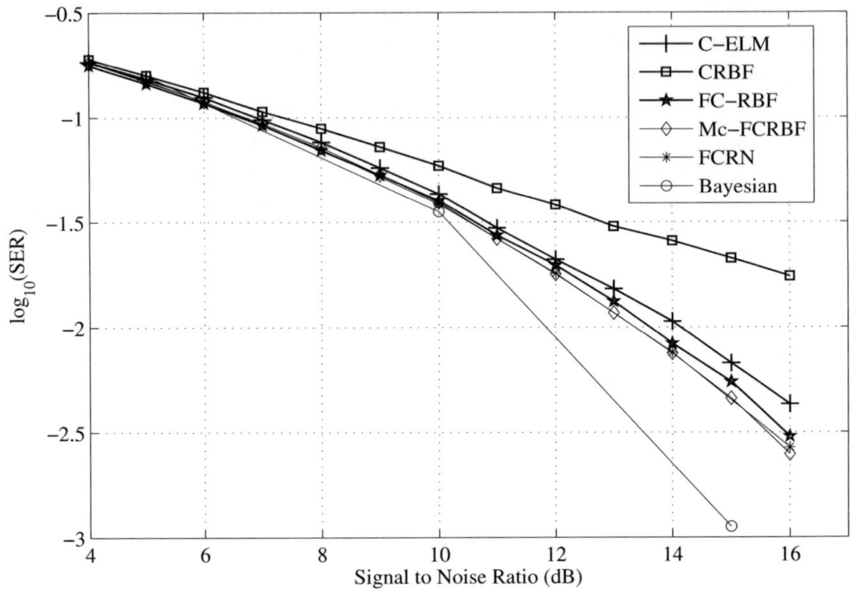

Fig. 5.3 Error Probability Curve for the Complex-valued RBF Based Equalizers

5.2.3 Adaptive Beam-Forming Problem

The demand for wireless mobile communication services is growing at a higher rate. However, the number of users for mobile and internet services is limited by the bandwidth of the wireless frequency spectrum. Increasing the channel capacity to accommodate many users without causing interference is a challenging task. Adaptive antennas, also referred as smart antennas is rapidly emerging as a solution to increase the system capacity within the limited spectrum to reduce interference as well as to enhance the overall wireless communication system performance. The beam pattern of an adaptive array can be adjusted to direct nulls in the direction of interference and main beam pointing to the desired signal directions as the adaptive array can differentiate the desired signals and unwanted co-channel interferences based on their different locations. The two basic parts of design for an adaptive antenna array are the direction of arrival estimation and the beam-forming. Adaptive beam-forming is a process in which the antenna array adapts its weights to the changing interference environment such that the antenna will receive only the desired signal and reject the interference and noise. It has found wide applications in a large diverse areas suc as radar, sonar, seismology, radio astronomy, speech, and biomedicine, besides wireless communications.

Over decades, many algorithms have been developed for this adaptive processes. They consist of the basic Sample Matrix Inversion algorithm (SMI) [12], unconstrained as well as constrained Least Mean Square (LMS) algorithm [13], structured gradient algorithm [14], recursive least squares method [12], constant modulus algorithm [15], conjugate gradient method [16] and neural network approach [17]. The beam-forming problem can be solved exactly for beam- pointing and null-steering by the SMI method. However, in practical cases when dealing with a large array, the matrix inversion is computationally expensive . Also, when the direction of arrival is not known, this method is of little use [18]. Widrow etal [13] showed that the variable weights of a signal processor can be automatically adjusted by a simple adaptive technique based on the least mean squares algorithm. But, the least mean squares algorithm uses a noisy estimate of the required gradient to adaptively estimate the weights of an optimal antenna array. Hence the estimation of weights is not accurate. In [14], a structured gradient algorithm, which used a structured estimate of the array co-relation matrix to estimate the gradient was suggested. However, it was later shown in [19] that the the covariance of the gradient estimated by the structured method is less sensitive to the look direction signal than that estimated by the standard method. A recursive least mean square algorithm and an improved LMS algorithm was developed in [20] and it was shown that while the recursive LMS algorithm is applicable for an array of arbitrary geometry, the improved LMS algorithm is useful for a linear array of equi-spaced elements.

Recently, neural networks have been widely employed for adaptive beam-forming problem and the focus of these works has been in the selection of an appropriate neural network along with a proper activation functions for handling the complex signals.

In [17], [21], a three-layer radial basis function network is used in the computation of the optimum weights of the fast tracking system which is used for constantly tracking the users. The radiation pattern of the antenna is then adapted to direct multiple narrow beams to the desired users and null the interfering sources. In [22], a comprehensive and detailed overview of the beam-forming problem along with different adaptive algorithms to adjust the required weights on the antennas have been presented. A complete overview of the neural network methods used in solving the adaptive beam-forming problem is given in [23]. In all these neural network methods, real-valued networks have been used, while, as will be shown in section 5.2.3.1, the beam-former has to deal inherently with complex-valued signals. Moreover, it has been shown in literature that fully complex-valued neural networks preserve the magnitude and phase information of complex-valued signals [8, 8, 24, 10, 25] and are hence, very efficient in approximating the complex-valued signals, compared to split complex-valued networks. Fully complex-valued multi-layer perceptron networks have been employed for adaptive beam-forming in [18], where it is shown that the beam-former based on FC-MLP steer the nulls better than the complex-valued least mean square method. In [26], the complex-valued radial basis function network is used for nonlinear beam-forming in multiple antenna aided communication systems that employ complex-valued quadrature phase shift keying modulation scheme.

5.2.3.1 The Beam-Forming Problem Formulation

Consider a coherent wireless communication system with M single-transmit-antenna users of the same carrier frequency ω and a receiver equipped with a linear antenna array consisting of N uniformly spaced elements. Fig. 5.4 gives the typical structure of a uniform linear array of sensor elements. The inter-element spacing, d, is equal to half the wavelength of the signal received in the senor array. Let θ be the angle of incidence that an incoming signal makes with the receiver array broadside.

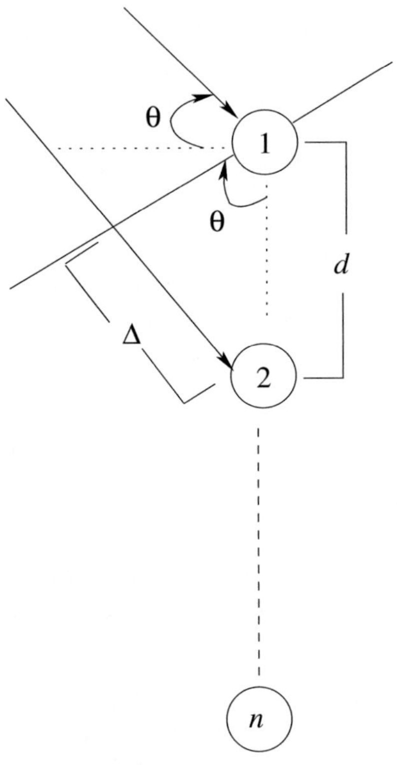

Fig. 5.4 The Sensor Array

The signal travels an additional distance of Δ between successive elements, i.e., for example, if the signal impinges on element 1, travelling a distance of D, it travels a distance of $D+\Delta$ to reach element 2. It can easily derived from the figure and with basic trigonometric identities that

$$\Delta = d \, sin \, \theta \tag{5.11}$$

Hence, the expression for time delay between two successive elements can be derived as

$$\delta_1 = \frac{d \sin \theta}{f \lambda}$$

and the time delay between the reference element and the nth element can be derived as

$$\delta_n = \frac{n}{f \lambda} d \sin \theta \tag{5.12}$$

Assuming the transmitted signal can be represented by a complex-valued sinusoid, and modulated at a frequency ω, the signal impinging at the array element n can be expressed as

$$\overline{z}_n(t) = e^{-j \, \omega \, (t \, - \, \delta_n)} \tag{5.13}$$

Demodulating this signal will give

$$\overline{z}_n = e^{j \, \omega \, \delta_n} \tag{5.14}$$

Replacing for δ_n from eq. (5.12) and simplifying, the signal received at the nth element of the receiver antenna array is given by eq. (5.15),

$$\overline{z}_n = e^{\frac{j \, 2 \, \pi \, n}{\lambda} \, d \, sin\theta} \tag{5.15}$$

Therefore, signals induced at the n receiver array elements due to the m sources will be

$$\mathbf{z} = [\overline{z}_0 \;\; \overline{z}_1 \;\; \overline{z}_2 \ldots \;\; \overline{z}_n]^T \tag{5.16}$$

Considering that a noise of η_n is added to signals impinging on each array element,

$$\mathbf{z} = [\overline{z}_0 + \eta_0 \;\; \overline{z}_1 + \eta_1 \;\; \overline{z}_2 + \eta_2 \ldots \;\; \overline{z}_n + \eta_n]^T \tag{5.17}$$

is the input to the beam-former. In eqs. (5.16) and (5.17), the superscript T represents the transpose of the vector.

Let w_k is the weight of the kth sensor, $\mathbf{w}_{bf}$ be the array of weights such that, $\mathbf{w}_{bf} = [w_1 w_2 \ldots \;\; w_k]^T$ and $\mathbf{w}_{bf}$ be the gain of the beam-former array for the signal from source m.

$$\mathbf{g} = \sum_{k=1}^{n} x_k \, w_k = \mathbf{z}^T \, \mathbf{w}_{bf} \tag{5.18}$$

w_k can now be computed by setting $g = 0$ for null-steering and $g = 1$ for beam pointing. The optimum weight vector $\mathbf{w}_{bf}$ would, now, be:

$$\mathbf{w}_{bf} = \mathbf{z}^{-1} \, \mathbf{g} \tag{5.19}$$

This analytical method of computing the optimum weight is known as the matrix method [12]. The matrix method is the exact method to estimate the weights of the beam-former, hence the beams and nulls can be steered in the desired directions. However, the computational complexity [27] of this method in dealing with large arrays, calls for a dynamic adaptive method. Therefore neural networks, being more adaptive [23], are employed to estimate the weights $\mathbf{w}_{bf}$. The objective is to train the neural network, such that the weights $\mathbf{w}_{bf}$ of eq. (5.20) are estimated.

In the adaptive method, there is a reference signal to be learnt by the adaptive processor, through adjustment of the weighting factor $\mathbf{w}_{bf}$. Here, $\mathbf{y}$ is the signal transmitted, and $\mathbf{z}$ is the signal received at the sensor array, given by eq. (5.17).

$$\mathbf{y} = \mathbf{w}_{bf}^{H}\,\mathbf{z} \tag{5.20}$$

It should be noted that the signals involved in an adaptive beam-forming problem are complex-valued. Hence, the fully complex-valued neural networks are a better choice to ensure accurate magnitude and phase approximation. In this section, we study the beam-forming performances of the complex-valued learning algorithms considered in this study. The performances of these algorithms are evaluated on the 5-antenna array configuration [18].

Array Configuration: A uniform linear array of 5 sensors is considered [18]. The array is trained to look at the desired signals coming from $-30°$ and $+30°$, and to suppress interferences from $-15°$, $0°$ and $+15°$ directions.

- Desired signal angles: $-30°$ and $+30°$
- Angles of nulls: $-15°, 0°$ and $+15°$

Dataset: The input to the neural network is given by eq. (5.17), with an additive Gaussian noise of 50dB SNR. Training dataset consists of 250 randomly chosen samples (50 for each signal/interference angles). The results were then compared with the beam pattern of the optimum matrix method.

Initial beampattern and the beampattern by matrix method: Fig 5.5(a) shows the beampattern before beam-forming and Fig. 5.5(b) gives the beam pattern using matrix method [12] for $\{-30°, +30°\}$ beams-pointing and $\{-15°, 0°, +15°\}$ nulls-steering. From the figure, it can be observed that signal power was the least at the nulls at $\{-15°, 0°, +15°\}$, thus the matrix method [12] provides excellent null suppression and beam pointing characteristics.

5.2.3.2 Complex-valued MLP Beam-Formers

In this section, we present the performance of the complex-valued MLP networks and learning algorithms, viz., SC-MLP, FC-MLP and IC-MLP in performing beam-forming of the adaptive array configuration discussed in section 5.2.3.1. In Table 5.5, the signal power of the different complex-valued MLP beam-formers, at the various signal angles ($-30°$ and $+30°$) and nulls ($-15°$, $0°$ and $+15°$) are presented. In all these algorithms, the network structure was fixed at 5 hidden neurons, before training. This was based on the heuristic procedure that is presented in Section 5.1 [6].

It can be observed from the table that the signal power of FC-MLP beam-former was the lowest at the nulls. In other words, FC-MLP beam-former did better suppression of nulls, compared to SC-MLP and IC-MLP beam-formers. However, the

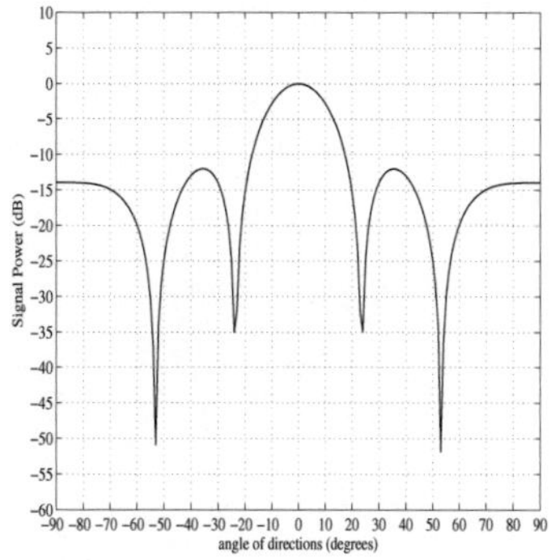

(a) Initial Beampattern: Beampattern of the Five Sensor Array

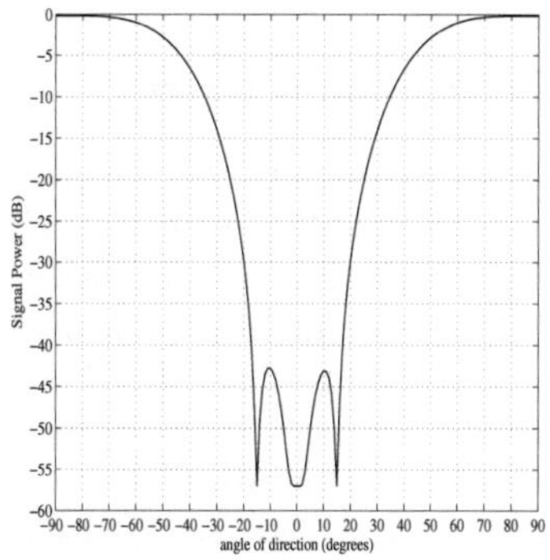

(b) Beampattern using the Matrix Method

Fig. 5.5 Initial Beampattern and Beampattern Using Matrix Method

signal power at nulls of IC-MLP was only about 1dB greater than that of FC-MLP beam-former. But, the signal power of the SC-MLP beam-former was at least 3dB greater than that of FC-MLP beam-former. Further, considering the signal power at the desired signal angles (i.e., the beampointing ability of the MLP beam-formers), the performance of IC-MLP beam-former is the best, though the SC-MLP and FC-MLP beam-formers are short of IC-MLP beam-former signal power only by 0.1dB. Hence, it can be inferred that both FC-MLP and IC-MLP are better choices for beam-formers, among the complex-valued MLP beam-formers.

In Fig. 5.6, the beam-forming performances of the various MLP based beam-formers are presented. It can be observed from the fig. that all the three MLP based beam-formers performed better suppression of nulls and direction of beams at desired angles. The signal power at other directions/angles are very similar to that of the signal power of the matrix method beam-former, as seen from Fig. 5.5(b). Thus,

Table 5.5 Performance comparison of CMLP beam-formers

Direction of arrival	Signal Power(dB)			
	SC-MLP	FC-MLP	IC-MLP	MM*
Beam-1:-30°	-13.98	-13.97	-13.87	-13.98
Null-1:-15°	-53.99	-57.2	-55.4	-57.02
Null-2:0°	-53.99	-57.34	-56.99	-57
Null-3:15°	-53.99	-57.33	-56	-57.02
Beam-2:30°	-13.98	-13.98	-13.86	-13.98

* Matrix Method

it can be deduced that all the complex-valued MLP beam-formers perform good beampointing and null suppression, though FC-MLP and IC-MLP beam-formers are better than SC-MLP beam-formers.

5.2.3.3 Complex-valued RBF Beam-Formers

In this section, we present the performance of various complex-valued RBF networks discussed in chapter 4, viz., CRBF with Gaussian activation function, C-ELM with Gaussian activation function and FC-RBF with the fully complex-valued "*sech*" activation function. In Table 5.6, the signal power/gain at different signal angles and nulls of the complex-valued RBF based beam-formers are tabulated. The signal power of the various CRBF beam-formers are also compared against the optimum matrix method beam-former. For the three networks shown, the number of hidden neurons was chosen to be 5, based on the heuristic procedure presented in Section 5.1 [6].

It can be observed from the table that the signal power of CRBF beam-former at the nulls is too large. Though the signal power of C-ELM beam-former at the nulls is quite less, it is still greater than -50dB. However, FC-RBF based beam-former outperforms both the CEBF and C-ELM (RBF) beam-former in the isolation of nulls and directing the beams in the desired direction. Signal power at the nulls are lesser than those of CRBF and C-ELM (RBF) based beam-formers. Although at a signal angle of $+30°$, the C-ELM performs better beam-pointing than FC-RBF beam-former, FC-RBF falls short of C-ELM only by 0.3dB. However at a signal angle of $-30°$, the signal power of FC-RBF beam-former is highest.

Comparing the performances of the complex-valued RBF based beam-formers, Mc-FCRBF and FCRN beam-formers perform excellent null suppression. Although the signal power at the desired angles of $-30°$ and $+30°$ of the Mc-FCRBF and FCRN beam-formers are nearly $3dB$ lesser than that of the matrix method based beam-former, its signal power at the nulls are lesser than the matrix method based beam-former, which is highly desirable. Hence, it can be inferred that, of the complex-valued RBF learning algorithms available in literature, Mc-FCRBF and FCRN are good in their beam-forming performance.

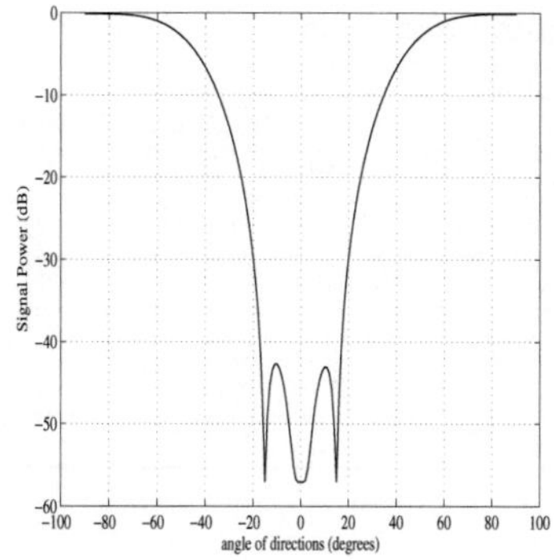

(a) Beampattern using SC-MLP Beamformer

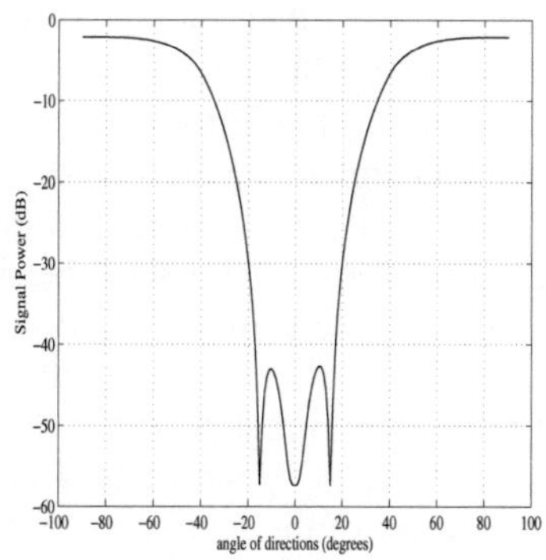

(b) Beampattern using FC-MLP Beamformer

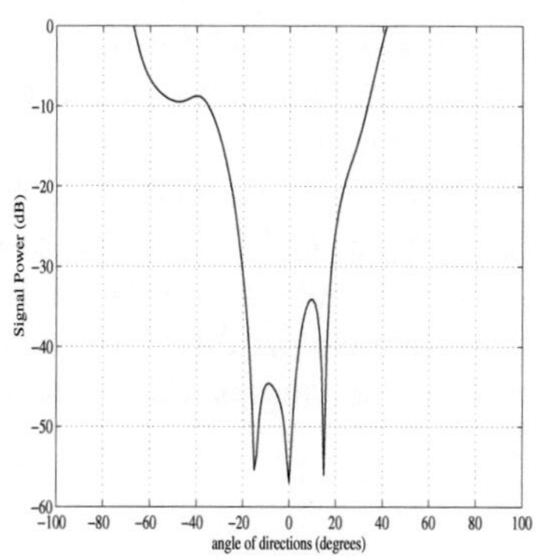

(c) Beampattern using IC-MLP Beamformer

Fig. 5.6 Beampattern Using Complex-valued MLP Based Beam-formers

Table 5.6 Performance comparison of CRBF beam-formers

Direction of arrival	Signal Power(dB)					
	CRBF	C-ELM	FC-RBF	Mc-FCRBF	FCRN	MM*
Beam-1:-30°	-17.94	-18.1	-16.99	-16.98	-14.27	-13.98
Null-1:-15°	-27.53	-48.29	-58.45	-60.54	-59.68	-57.02
Null-2:0°	-27	-41.6	-57.23	-59.6	-60.4	-57
Null-3:15°	-28.33	-47.5	-56.32	-59.6	-59.68	-57.02
Beam-2:30°	-17.92	-16.7	-17.00	-16.9	-14.1	-13.98

* Matrix Method

In Fig. 5.7, the beam-forming of various complex-valued RBF beam-formers are presented.

From the figure, it can be observed that the signal power of the CRBF beam-former at the signal angles and nulls differ only by 10dB. Hence, the CRBF beam-former is not efficient in suppressing the nulls and in directing the beams to the desired direction. Besides, at other signal angles, too, the signal power of the CRBF beam-former is not similar to that of the beampattern generated by the optimum matrix method. Hence, it can be inferred that CRBF beam-formers are not a good choice to suppress nulls and direct beams in desired directions in a beam-forming application. From the fig. 5.7, it can also be observed that C-ELM beam-former provides better isolation of nulls. The signal power at the nulls and desired signals are very distinct. However, the signal power at nulls are greater than -50dB, which is atleast 7dB greater than the power of the matrix method beam-former.

From Fig. 5.7, it can be observed that the beampattern of FC-RBF, FCRN and Mc-FCRBF beam-formers are very similar to the optimum matrix method beam-former. The signal power at nulls are very similar to that of the optimum matrix method beam-former, while at the desired signal angles, the signal power is 3dB lesser than that of the matrix method beam-former. However, it provides better null suppression and beampointing, compared to the CRBF and C-ELM beam-formers. Hence, it can be inferred that the FC-RBF beam-former performs better than the other complex-valued RBF beam-formers in the beam-forming application. Similarly, comparing the performance of FC-RBF beam-former to that of the Mc-FCRBF beam-former, the Mc-FCRBF beam-former has better null suppression performance, which is even better than that of the matrix method based beam-former, although, its signal power at the signal angles are a little less than that of the matrix method based beam-former.

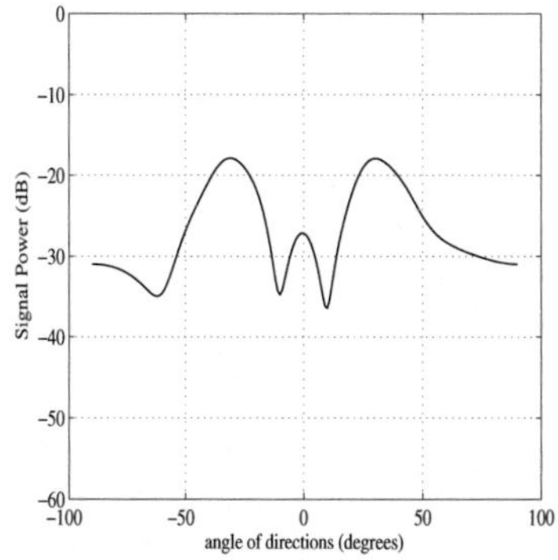

(a) Beampattern using CRBF Beam-former

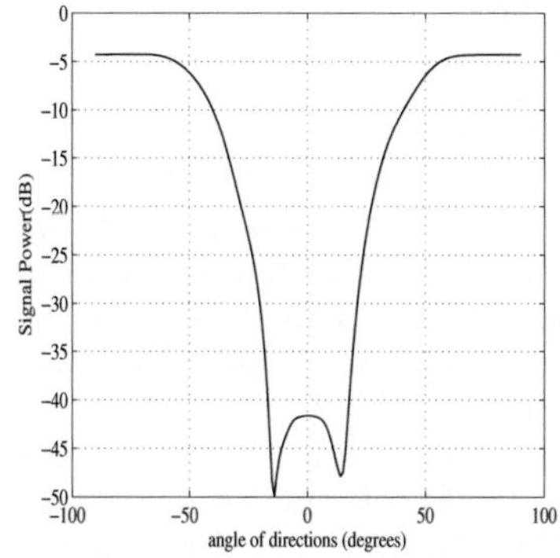

(b) Beampattern using C-ELM (RBF) Beam-former

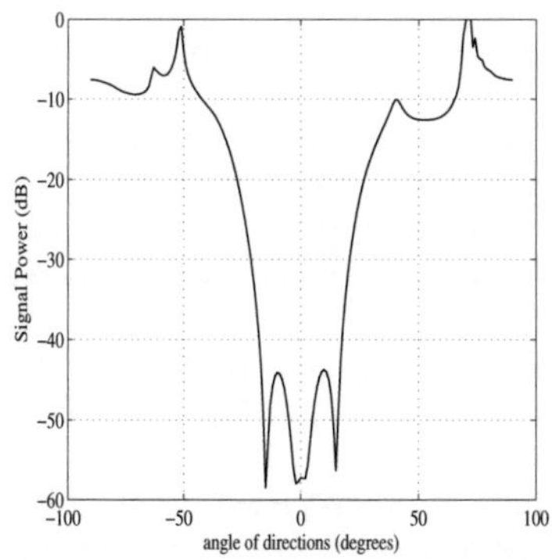

(c) Beampattern using FC-RBF Beam-former

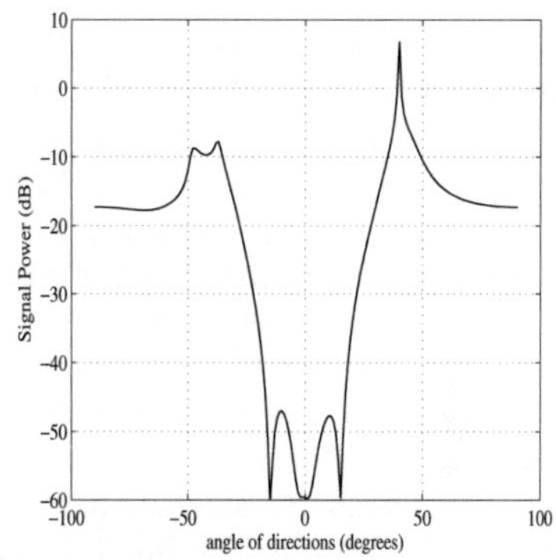

(d) Beampattern using Mc-FCRBF Beam-former

Fig. 5.7 Beampattern Using Complex-valued RBF Based Beam-formers

5.3 Summary

In this chapter, the function approximation performances of FC-MLP, IC-MLP, FC-RBF, Mc-FCRBF and FCRN algorithms have been studied in comaprison with CRBF and C-ELM learning algorithms on two synthetic complex-valued function approximation problems and two practical problems consisting of a QAM channel equalization problem and an adaptive beam-forming problem. Performance results show that the phase approximation ability of IC-MLP is better than that of FC-MLP by at least 60% with better magnitude approximation. Similarly, the approximation ability of FC-RBF is better than C-RBF and C-ELM by at least 25%, which is further improved by the meta-cognitive component of Mc-FCRBF. It is also observed that FCRN performs better approximation of both the mangitude and phase of the complex-valued signals with reduced computational effort.

References

1. Proakis, J.G., Salehi, M.: Digital Communication. McGraw-Hill Higher Education, New York (2008)
2. Patra, J.C., Pal, R.N., Baliarsingh, R., Panda, G.: Nonlinear channel equalization for QAM constellation using artificial neural networks. IEEE Transactions on System, Man and Cybernetics, Part B: Cybernetics 29(2), 262–271 (1999)
3. Chen, S., McLaughlin, S., Mulgrew, B.: Complex valued radial basis function network, part II: Application to digital communications channel equalization. Signal Processing 36(2), 175–188 (1994)
4. Cha, I., Kassam, S.A.: Channel equalization using adaptive complex radial basis function networks. IEEE Journal on Selected Areas in Communications 13(1), 122–131 (1995)
5. Li, M.B., Huang, G.-B., Saratchandran, P., Sundararajan, N.: Fully complex extreme learning machine. Neurocomputing 68(1-4), 306–314 (2005)
6. Suresh, S., Omkar, S.N., Mani, V., Prakash, T.N.G.: Lift coefficient prediction at high angle of attack using recurrent neural network. Aerospace Science and Technology 7(8), 595–602 (2003)
7. Leung, H., Haykin, S.: The complex backpropagation algorithm. IEEE Transactions on Signal Processing 39(9), 2101–2104 (1991)
8. Kim, T., Adali, T.: Fully complex multi-layer perceptron network for nonlinear signal processing. Journal of VLSI Signal Processing 32(1/2), 29–43 (2002)
9. Savitha, R., Suresh, S., Sundararajan, N., Saratchandran, P.: A new learning algorithm with logarithmic performance index for complex-valued neural networks. Neurocomputing 72(16-18), 3771–3781 (2009)
10. Chen, S., McLaughlin, S., Mulgrew, B.: Complex valued radial basis function network, part I: Network architecture and learning algorithms. EURASIP Signal Processing Journal 35(1), 19–31 (1994)
11. Savitha, R., Suresh, S., Sundararajan, N.: A fully complex-valued radial basis function network and its learning algorithm. International Journal of Neural Systems 19(4), 253–267 (2009)
12. Trees, H.L.V.: Optimum Array Processing, Detection, Estimation and Modulation Theory: Part IV. John Wiley and Sons, New York (2001)
13. Widrow, B., Mantey, P.E., Griffiths, L.J., Goode, B.B.: Adaptive antenna systems. Proceedings of the IEEE 55(12), 2143–2159 (1967)
14. Godara, L.C., Gray, D.A.: A structured gradient algorithm for adaptive beamforming. Journal of the Acoustic Society of America 86(3), 1040–1046 (1989)
15. Papadias, C.B., Paulraj, A.: A constant modulus algorithm for multiuser signal separation in presence of delay spread using antenna arrays. IEEE Signal Processing Letters 4(6), 178–181 (1997)
16. Mandyam, G.D., Ahmed, N., Srinath, M.D.: Adaptive beamforming based on the conjugate gradient algorithm. IEEE Transactions on Aerospace and Electronic Systems 33(1), 343–347 (1997)
17. El Zooghby, A.H., Christodoulou, C.G., Georgiopoulos, M.: Neural network based adaptive beamforming for one- and two-dimensional antenna arrays. IEEE Transactions on Antennas Propagation 46(12), 1891–1893 (1998)
18. Suksmono, A.B., Hirose, A.: Intelligent beamforming by using a complex-valued neural network. Journal of Intelligent and Fuzzy Systems 15(3-4), 139–147 (2004)
19. Godara, L.C.: Performance analysis of structured gradient algorithm antenna array processing. IEEE Transactions on Antennas Propagation 38(7), 1078–1083 (1990)

20. Godara, L.C.: Improved LMS algorithm for adaptive beamforming. IEEE Transactions on Antennas Propagation 38(10), 1631–1635 (1990)
21. Song, X., Wang, J., Niu, X.: Robust adaptive beamforming algorithm based on neural network. In: IEEE International Conference on Automation and Logistics (ICAL 2008), pp. 1844–1849 (2008)
22. Godara, L.C.: Application of antenna arrays to mobile communications, part II: Beamforming and direction-of-arrival considerations. Proceedings of the IEEE 85(8), 1195–1245 (1997)
23. Du, K.L., Lai, A.K.Y., Cheng, K.K.M., Swamy, M.N.S.: Neural methods for antenna array signal processing: A review. Signal Processing 82(4), 547–561 (2002)
24. Savitha, R., Suresh, S., Sundararajan, N., Saratchandran, P.: Complex-valued function approximation using an improved BP learning algorithm for feed-forward networks. In: IEEE International Joint Conference on Neural Networks (IJCNN 2008), June 1-8, pp. 2251–2258 (2008)
25. Savitha, R., Suresh, S., Sundararajan, N.: Complex-valued function approximation using a fully complex-valued RBF (FC-RBF) learning algorithm. In: International Joint Conference on Neural Networks (IJCNN 2009), pp. 2819–2825 (2009)
26. Chen, S., Hong, X., Harris, C.J., Hanzo, L.: Fully complex-valued radial basis function networks: Orthogonal least squares regression and classification. Neurocomputing 71(16-18), 3421–3433 (2008)
27. Monzingo, R.A., Miller, T.W.: Introduction to Adaptive Arrays. SciTech. Publishing, Raleigh (2004)

Chapter 6
Circular Complex-valued Extreme Learning Machine Classifier

Artificial Neural Networks (ANN) were originally inspired by the central nervous system and its components that constitute the biological neural network, as investigated by the Neuroscience community. Ever since then, several tasks of human activity have been emulated by the ANNs. Classification is one such decision making task that occurs frequently in human activity and one that has been emulated in the artificial neural network framework. A classification task in the ANN framework is defined as assigning an object to a predefined group or class based on a set of object attributes. As the ANNs are capable of constructing complex decision boundaries without any assumption on the statistics on the input data, they have been used to perform classification tasks in a large range of applications spanning from business to medical diagnosis to speech recognition. Over the past twenty years, supervised learning has become a standard tool of choice to analyze the data in many fields, particularly in classfication problems. The objective of the classification problem is to approximate the decision surface described by the training data and predict the class label of the unknown data as accurately as possible. Recently, it has been shown by Nitta that the complex-valued neural networks have better computational power than real-valued networks [1] and they outperform real-valued networks in their ability to approximate the decision boundaries to solve classification problems. Moreover, it has been shown by Nitta [2, 3] that a fully complex-valued neural network with a split type of activation function has two decision boundaries that are orthogonal to each other. These decision boundaries help the complex-valued neural network to perform classification tasks more efficiently than real-valued networks. These findings have inspired researchers to develop efficient classifiers in the Complex domain. The 'Multi Layered neural network based on Multi Valued Neurons (MLMVN)' [4] and the single-layered network with phase encoded transformation, referred to here as, 'Phase Encoded Complex Valued Neural Network (PE-CVNN)' [5] are the two such complex-valued classifiers available in the literature. In this chapter, we briefly discuss these classifiers and show that the complex-valued learning algorithms described in the previous chapters can be modified to solve real-valued classification problems. In addition, we present an efficient and fast learning

S. Suresh et al.: Supervised Learning with Complex-valued Neural Networks, SCI 421, pp. 109–123.
springerlink.com

complex-valued classifier, referred to as 'Circular Complex-valued Extreme Learning Machine (CC-ELM)' to solve real-valued classification problems. CC-ELM is a single hidden layer network with a non-linear input and hidden layer and a linear output layer. A circular transformation with a translational/rotational bias that performs an unique one-to-one transformation of the real-valued feature to the Complex plane is used as an activation function for the neurons in the input layer. Neurons in the hidden layer employ a fully complex-valued Gaussian-like ('*sech*') activation function. The input parameters of the CC-ELM are chosen randomly and the output weights are computed analytically. This chapter also presents an analytical proof to show that the decision boundaries of a single complex-valued neuron at the hidden and output layer of the CC-ELM consists of two hyper-surfaces that intersect orthogonally.

6.1 Complex-valued Classifiers in the Literature

In this section, we review the existing complex-valued classifiers. First, we define the real-valued classification problem in the Complex domain and describe the learning algorithms of the two complex-valued classifiers, viz., the Multi-Layer neural network based on Multi-Valued Neurons (MLMVN) and the Phase Encoded Complex-valued Neural Network (PE-CVNN).

6.1.1 Description of a Real-valued Classification Problem Done in the Complex Domain

Suppose we have N observations $\{(\mathbf{x}_1, c_1), \cdots, (\mathbf{x}_t, c_t), \cdots, (\mathbf{x}_N, c_N)\}$, where $\mathbf{x}_t \in \Re^m$ be the m-dimensional input features of t-th observation, $c_t \in [1, 2, \cdots, n]$ are its class labels, and n is the number of distinct classes. The observation data $(\mathbf{x}_t, c_t)$ are random in nature and the observation $\mathbf{x}_t$ provides some useful information on the probability distribution over the observation data to predict the corresponding class label (c_t) with a certain accuracy.

To solve the real-valued classification problems using complex-valued neural networks, the coded class label ($\mathbf{y}^t = [y_1^t \cdots y_k^t \cdots y_n^t]^T$) is defined in the Complex domain as:

$$y_k^t = \begin{cases} 1 + i1 & \text{if} \quad c_t = k \\ -1 - i1 & \text{otherwise} \end{cases} \quad k = 1, 2, \cdots, n \tag{6.1}$$

The real-valued classification problem using complex-valued neural networks can be viewed as finding the decision function F that maps the real-valued input features to the complex-valued coded class labels, i.e., $F : \Re^m \to \mathbb{C}^n$. For notational convenience, the superscript t will be dropped in the rest of the chapter.

6.1.2 Multi-Layer Neural Network Based on Multi-Valued Neurons (MLMVN)

The complex-valued MLMVN using multi-valued neurons was developed in [4]. A single multi-valued neuron maps m inputs and a single output. The mapping is described by a multi-valued function of m variables ($f_m(x_1, \cdots, x_m)$) with $m+1$ complex-valued weights as the parameters, and is given by

$$f_m(x_1, x_2, \cdots, x_m) = P(v_0 + v_1 x_1 + \cdots + v_m x_m) \tag{6.2}$$

where $P(.)$ is the activation function of the neuron, given by

$$P(z) = exp(i(arg(z))) = \frac{z}{|z|} \tag{6.3}$$

where $z = v_0 + v_1 x_1 + \cdots + v_m x_m$ is the weighted sum, and $|z|$ is the modulo of the complex number z. The function defined in Eq. (6.3) maps the Complex plane into the whole unit circle and is continuous. It must also be noted that the function is not differentiable as a function of a complex-valued variable.

However, as the learning in the MLMVN is reduced to the movement along the unit circle, the derivative of the activation function is not required because it is impossible to move in an incorrect direction. Any direction of movement along the circle will lead to the target and the shortest way of this movement is determined by the error which is the difference between the desired output and the actual output. The weight update rule for a multi-valued neuron used in an MLMVN is given by:

$$V_{m+1} = V_m + \frac{\eta_v}{m+1} \tilde{\delta} \overline{X} \tag{6.4}$$

$$\text{where } \tilde{\delta} = \frac{1}{|z|}\left(y - \frac{z}{|z|}\right) \tag{6.5}$$

From Eq. (6.4), it can be observed that using this approach, the function is a smooth function of the weights. Moreover, a small change in the inputs or weights does not result in a significant change of z. However, as the input features are mapped on to a full unit circle, this mapping results in the same complex-valued features for real-valued features with a value of both 0 and 1. Hence, this transformation is not unique. In addition, the multi-valued neurons map the complex-valued inputs to C discrete outputs on the unit circle. Thus, as the number of classes (C) increases, the number of sectors with the unit circle increases. As a result, the region of each sector (representing each class) within the unit circle decreases, increasing the chances of misclassification. Furthermore, the output neurons of the MLMVN have multiple discrete values and hence, the MLMVN classifier lacks the orthogonal decision boundaries that are the main characteristics of complex-valued neural classifiers.

As the transformation used in the MLMVN does not perform an one-to-one mapping of the real-valued features onto the Complex plane, Amin et. al. [5] proposed a phase encoded transformation to convert the real-valued features to the complex domain. They also developed a single layered complex-valued classifier using this phase encoded transformation. We describe the transformation and the learning algorithm of the network developed in the next section.

6.1.3 Phase Encoded Complex-Valued Neural Network (PE-CVNN)

In PE-CVNN [5], the following phase encoded transformation is used to transform the real-valued input features to the complex domain:

$$\text{Let } x_j \in [a,b], \text{where} a,b \in \Re, \text{ then } \phi = \frac{\pi(x_j - a)}{b-a} \tag{6.6}$$

$$\text{and } z_j = e^{i\phi} = cos(\phi) + isin\phi \tag{6.7}$$

It can be observed from Eq. (6.6) that the the transformation linearly maps the input features from $x \in [a,b] \rightarrow \phi \in [0,\pi]$.

The weighted inputs at the output layer of the PE-CVNN are given by:

$$z_{ok} = z_{ok}^R + iz_{ok}^I = \sum_{j=1}^{m} v_{kj} z_j + \theta_k \tag{6.8}$$

and the output of the network is

$$\widehat{y}_k = f_{\mathbb{C}\rightarrow\Re} \tag{6.9}$$

$$\text{and } f_{\mathbb{C}\rightarrow\Re} = \sqrt{(f_R(z_{ok}^R))^2 + (f_R(z_{ok}^I))^2} \tag{6.10}$$

$$\text{OR } f_{\mathbb{C}\rightarrow\Re} = (f_R(z_{ok}^R) - f_R(z_{ok}^R))^2$$

$$\text{where, } f_R(z_{ok}^R) = 1/1 + exp(-(z_{ok}^R)) \text{ and}$$

$$f_R'(z_{ok}^R) = f_R(z_{ok}^R)(1 - f_R(z_{ok}^R)) \tag{6.11}$$

The residual error of the network output is given as:

$$e_k = y_k - \widehat{y}_k \tag{6.12}$$

The PE-CVNN uses a gradient descent based learning rule with the mean squared error criterion to update the weights of the network. The weight update rule is given by:

$$\Delta v_{kj} = \overline{z_j} \Delta \theta_k$$

$$\text{where} \Delta \theta_k = \Delta \theta_k^R + i \Delta \theta_k^I \tag{6.13}$$

If the activation function used in the PE-CVNN is given by Eq. (6.11), then $\Delta \theta_k^R$ is

$$\Delta \theta_k^R = 2\eta e_k \left(f_R(z_{ok}^R) - f_R(z_{ok}^I)\right) f_R'(z_{ok}^R) \tag{6.14}$$

$$\text{and } \Delta \theta_k^I = 2\eta e_k \left(f_R(z_{ok}^I) - f_R(z_{ok}^R)\right) f_R'(z_{ok}^I) \tag{6.15}$$

The transformation used in the PE-CVNN considers only the first two quadrants of the Complex plane and does not fully exploit the advantages of the orthogonal decision boundaries offered by the complex-valued neural networks. In addition, the activation functions used in the PE-CVNN are similar to those used in the split-type complex-valued neural networks, as shown in Eqs. (6.10) and (6.11). Also, the gradients used in the PE-CVNN are not fully complex-valued [6] thereby resulting in a significant loss of complex-valued information. PE-CVNN uses a gradient-descent based batch learning algorithm that requires a significant computational effort to approximate the decision surface.

It is noteworthy that the transformations used in MLMVN and PE-CVNN can be used to transform the real-valued input features to the Complex domain and any of the complex-valued learning algorithms discussed in Chapters 2-4 can be used to solve real-valued classification problems using the transformed features. Two factors play a vital role in deciding the classification ability of the complex-valued neural networks:

1. The transformation used to convert the real-valued input features to the Complex domain
2. The activation function and learning algorithm used in the complex-valued neural network.

In the next section, we discuss the modifications required to enhance the decision making ability of FC-MLP, FC-RBF and Mc-FCRBF such that these algorithms can be used to solve real-valued classification problems.

6.1.4 Modifications in FC-MLP, FC-RBF and Mc-FCRBF Learning Algorithm to Solve Real-valued Classification Problems

The learning algorithm of FC-MLP, FC-RBF and Mc-FCRBF described in Chapters 2 and 3 have been used to solve complex-valued function approximation problems in Chapter 5. Although they can also be used to approximate the decision surface to solve real-valued classification problems, we modify these algorithms to improve their classification performance. In this respect, the mean squared error defined in Eq. (3.14) is replaced with the hinge loss function and the criteria for parameter update in Mc-FCRBF is modified to incorporate a classification measure also.

Hinge loss error function: Recently, it was shown in [7] and [8] that in real-valued classifiers, the hinge loss function helps the classifier to estimate the posterior probability more accurately than the mean squared error function. Hence, while using FC-MLP, FC-RBF and Mc-FCRBF to solve real-valued classification problems, we use the hinge loss error function defined as:

$$e_l^t = \begin{cases} 0 \text{ if } (y_l^{t^R})(\widehat{y}_l^{t^R}) > 0 \\ y_l^t - \widehat{y}_l^t \quad \text{otherwise} \end{cases} \quad l = 1, 2, \cdots n \tag{6.16}$$

where the superscript R refers to the real-part of the Complex signal.

Criteria for parameter update: In addition to the hinge loss error function, we modify the parameter update criteria of Mc-FCRBF to enhance its decision making ability. While solving real-valued classification problems in the Complex domain, it has to be ensured that the predicted class label of Mc-FCRBF ($\widehat{c}^t$) is the same as that of the target class label (c^t). Therefore, we have modified the parameter update conditions to accommodate this class label information also. Accordingly, the sample learning condition of Mc-FCRBF (Eq. (3.38)) to solve real-valued classification problems is:

$$\text{If } \widehat{c}^t \neq c^t \text{ OR } \left(M_t^e \geq E_l^M \text{ AND } \phi_t^e \geq E_l^\phi\right) \tag{6.17}$$

Then, update the network parameters according to Eqs. (3.31), (3.32), and (3.33). Here, $\widehat{c}^t$ is the predicted class label for the sample t defined as:

$$\widehat{c}^t = \max_{l=1,2,\cdots,C} \text{ real } \left(\widehat{y}_l^t\right) \tag{6.18}$$

6.2 Circular Complex-valued Extreme Learning Machine Classifier

In this section, a fast learning fully-complex valued classifier called 'Circular Complex-valued Extreme Learning Machine' is developed to solve real-valued classification tasks.

6.2.1 Architecture of the Classifier

The circular complex-valued extreme learning machine classifier is a single hidden layer network with m input neurons, h hidden neurons and n output neurons, as shown in Fig. 6.1. The neurons in the input layer employ a non-linear circular transformation as the activation function, as shown in the inset of Fig. 6.1.

The circular transformation that transforms the real-valued input features into the Complex domain ($\Re \rightarrow \mathbb{C}$) is given by

$$z_l = sin\left(ax_l + ibx_l + \alpha_l\right), \quad l = 1, 2, \cdots, m \tag{6.19}$$

where $a,\ b,\ \alpha_l \in \Re^+$ are non-zero constants and x_l is the input feature normalized in $[0,1]$. The scaling factors a, and b, and the translational, rotational bias term α_l are randomly chosen such that $0 < a, b \leq 1$, and $0 < \alpha_l < 2\pi$. The translational, rotational bias term α_l shifts the origin of the resultant complex-valued feature (z_l) and rotates it to any quadrant of the Complex domain.

The effect of the circular transformation for different values of the translational, rotational bias term (α_l) is shown in Fig. 6.2. It can be observed from this figure that the randomly chosen bias terms perform the translation/rotation of the feature vector in different quadrants of the complex-valued feature space. Therefore, the bias term ($\alpha_l,\ l = 1,\ \cdots,\ m$) associated with each input feature ensures that all the input

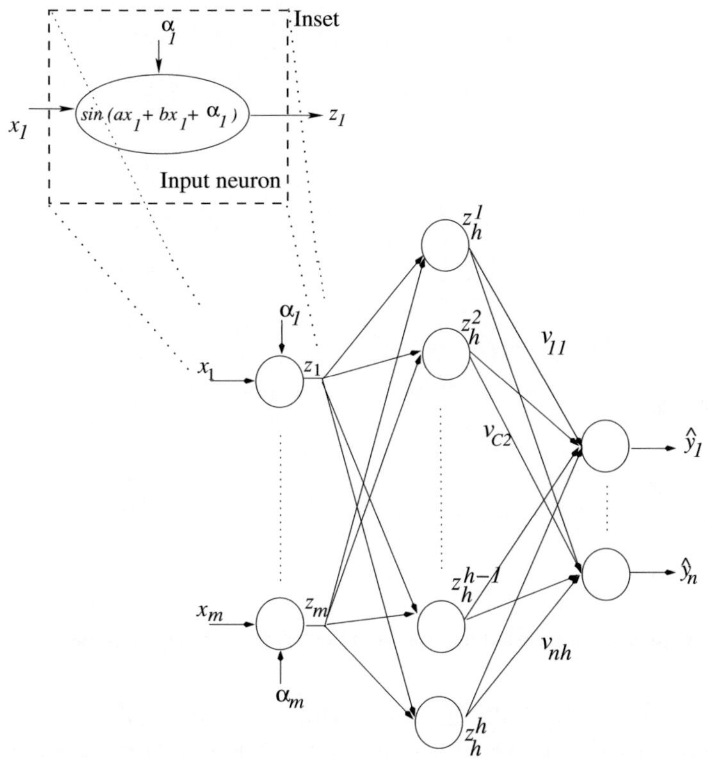

Fig. 6.1 The architecture of a circular complex-valued extreme learning machine. The expanded box (inset figure) shows the actual circular transformation function which maps real-valued input feature to complex-valued feature.

features are not mapped onto the same quadrant of the Complex plane. Thus, as the input features are well distributed in the Complex plane, the CC-ELM exploits the orthogonal decision boundaries of the fully complex-valued neural networks more efficiently.

The essential properties of a fully complex-valued activation function given in [6] states that for a complex-valued non-linear function to be used as an activation function, the function has to be analytic and bounded *almost everywhere*. As the circular transformation is used as an activation function in the input layer, we need to ensure that the transformation satisfies the essential properties of a fully complex-valued activation function. The '*sine*' activation function is an analytic function with an essential singularity at ∞. The net input to the '*sine*' function

$$ax_l + ibx_l + \alpha_l = \infty \text{ if } ax_l + \alpha_l = \infty \text{ or } bx_l = \infty \tag{6.20}$$

Since the transformation constants (a,b) and the translational/rotational bias (α_l) are restricted between $(0,1]$ and $(0,2\pi)$ respectively, the transformation becomes unbounded only when the input feature is ∞. Since, the real-valued features are

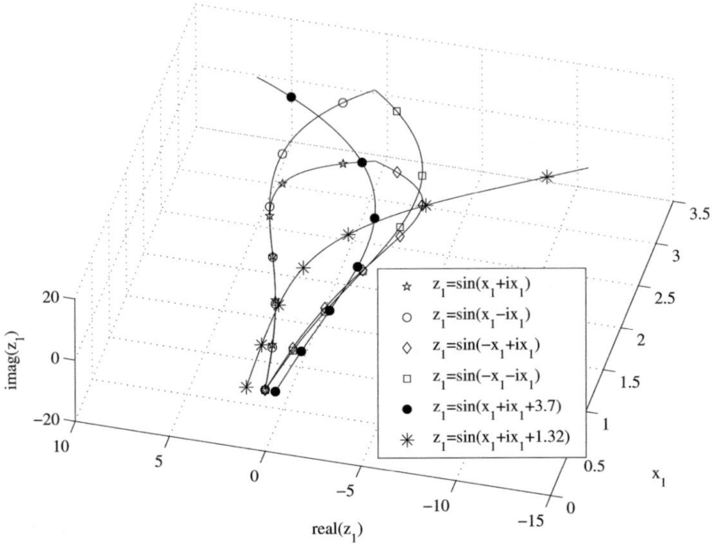

Fig. 6.2 Significance of translational/rotational bias term (α) in the circular transformation

normalized between $[0,1]$, the circular transformation is analytic and bounded *almost everywhere*. Hence, the circular transformation used in the input layer of the CC-ELM is a valid activation function.

The neurons in the hidden layer of the CC-ELM employ a fully-complex valued '*sech*' activation function (Gaussian like) as developed in [9]. The hidden layer response (z_h^j) of the CC-ELM is given by

$$z_h^j = sech\left[\boldsymbol{\sigma}_\mathbf{j}^\mathbf{T}\left(\mathbf{z}-\mathbf{c}_\mathbf{j}\right)\right], \quad j=1,2,\cdots,h \tag{6.21}$$

where $\boldsymbol{\sigma}_\mathbf{j} \in \mathbb{C}^\mathbf{m}$ is the complex-valued scaling factor of the j-th hidden neuron, $\mathbf{c}_j \in \mathbb{C}^m$ is the complex-valued center of the j-th hidden neuron and the superscript T represents the matrix transpose operator.

The neurons in the output layer employ a linear activation function. The output ($\widehat{\mathbf{y}} = [\widehat{y}_1 \cdots \widehat{y}_n \cdots \widehat{y}_n]^T$) of the CC-ELM network with h hidden neurons is

$$\widehat{y}_n = \sum_{j=1}^{h} v_{kj} z_h^j, \quad k=1,2,\cdots,n \tag{6.22}$$

where w_{kj} is the output weight connecting the j-th hidden neuron and the k-th output neuron.

The estimated class label ($\widehat{c}$) is then obtained using

$$\widehat{c} = arg \max_{n=1,2,\cdots,n} \text{real-part of } \widehat{y}_k \tag{6.23}$$

6.2.2 Learning Algorithm of CC-ELM

The output of CC-ELM given in Eq. (8.17) can be written in matrix form as,

$$\widehat{Y} = VH \tag{6.24}$$

where V is the matrix of all output weights connecting hidden and output neurons. The H is the response of hidden neurons for all training samples and is given as

$$H(V,U,Z) = \begin{bmatrix} sech\left(\boldsymbol{\sigma}_1^T \|\mathbf{z}_1 - \mathbf{c}_1\|\right) & \cdots & sech\left(\boldsymbol{\sigma}_1^T \|\mathbf{z}_N - \mathbf{c}_1\|\right) \\ \vdots & \vdots & \vdots \\ sech\left(\boldsymbol{\sigma}_j^T \|\mathbf{z}_1 - \mathbf{c}_j\|\right) & \cdots & sech\left(\boldsymbol{\sigma}_j^T \|\mathbf{z}_N - \mathbf{c}_j\|\right) \\ \vdots & \vdots & \vdots \\ sech\left(\boldsymbol{\sigma}_h^T \|\mathbf{z}_1 - \mathbf{c}_h\|\right) & \cdots & sech\left(\boldsymbol{\sigma}_h^T \|\mathbf{z}_N - \mathbf{c}_h\|\right) \end{bmatrix} \tag{6.25}$$

Note that H is $h \times N$ hidden layer output matrix. The j-th row of the H matrix represents the hidden neuron response (z_h^j) with respect to the inputs $\mathbf{z}_1, \cdots, \mathbf{z}_N$.

In CC-ELM, the transformation constants (a, b), the translational/rotation bias term ($\boldsymbol{\alpha}$), center (C) and scaling factor ($\boldsymbol{\sigma}$) are chosen randomly and the output weights V are estimated by the least squares method according to:

$$V = YH^{\dagger} \tag{6.26}$$

where $H^{\dagger}$ is the generalized Moore-Penrose pseudo-inverse [10] of the hidden layer output matrix and Y is the complex-valued coded class label.

In short, the CC-ELM algorithm can be summarized as:

- For a given training set (X, Y), select the appropriate number of hidden neurons h.
- Randomly select $0 < a$, b, $0 < \alpha_l < 2\pi$, the neuron scaling factor ($\boldsymbol{\sigma}$) and the neuron centers (C).
- Then calculate the output weights V analytically: $V = YH^{\dagger}$.

Performance of the CC-ELM is influenced by the selection of appropriate number of hidden neurons. Recently, an incremental constructive method to determine the appropriate number of hidden neurons for the C-ELM has been presented in [11]. In this paper, we use a simple neuron incremental-decremental strategy, similar to the one presented in [12] for real-valued networks. The following steps are followed to select the appropriate number of hidden neurons for the CC-ELM network:

Step 1. Select a network with a minimum configuration ($h = m + n$).
Step 2. Select the input weights randomly and compute the output weights analytically.
Step 3. Use leave-one cross-validation to determine training/validation accuracy from the training data.

Step 4. Increase h until the validation accuracy improves and return to Step 2.
Step 5. If the validation accuracy decreases as h increases, then stop.

From this section, it can be observed that the two important properties of the CC-ELM classifier are that:

- The CC-ELM classifier uses a unique circular transformation that makes an one-to-one mapping while transforming the real-valued input features to the Complex domain ($\Re \rightarrow \mathbb{C}$).
- The CC-ELM classifier requires lesser computational effort than other complex-valued classifiers as the weights are computed analytically.

6.2.3 Orthogonal Decision Boundaries in CC-ELM

In this section, we show that the three layered CC-ELM with the fully complex-valued *sech* activation function at the hidden layer exhibits orthogonal decision boundaries. Since CC-ELM maps the complex-valued input features to the higher dimensional Complex plane in the hidden layer randomly, we prove the existence of orthogonal decision boundaries in the output neuron with respect to the hidden layer output. Next, we also show that the decision boundaries formed by the real and imaginary parts of the hidden layer response with respect to the input are orthogonal to each other.

6.2.4 Case (i): Orthogonality of Decision Boundaries in the Output Layer

The responses of the kth output neuron can be written as:

$$\widehat{y}_k = \sum_{j=1}^{h} v_{kj} z_h^j; \; k = 1, \cdots, n \tag{6.27}$$

$$= \sum_{j=1}^{h} \left(v_{kj}^R + i v_{kj}^I\right) \left(z_h^{jR} + i z_h^{jI}\right) \tag{6.28}$$

where the superscripts R and I represents the real and imaginary parts of the complex-valued signals, respectively. Therefore,

$$\widehat{y}_k = \widehat{y}_k^R + i\widehat{y}_k^I = \sum_{j=1}^{h} \left(v_{kj}^R z_h^{jR} - v_{kj}^I z_h^{jI}\right) + i\left(v_{kj}^R z_h^{jI} + v_{kj}^I z_h^{jR}\right) \tag{6.29}$$

From the above equation, we can see that the real-part and the imaginary-part of the kth output neuron forms two decision boundaries with respect to the hidden layer output ($\mathbf{z_h}$). The two decision boundaries are

$$\widehat{y}_k^R = \sum_{j=1}^{h} \left(v_{kj}^R z_h^{jR} - v_{kj}^I z_h^{jI} \right) \rightarrow \mathbf{S}^R \tag{6.30}$$

$$\text{and } \widehat{y}_k^I = \sum_{j=1}^{h} \left(v_{kj}^R z_h^{jI} + v_{nj}^I z_h^{jR} \right) \rightarrow \mathbf{S}^I \tag{6.31}$$

The real-part decision boundary $\mathbf{S}^R$ classifies the hidden layer output $\mathbf{z}_h \in \mathbb{C}^h$ into two decision regions

$$\left\{ \mathbf{z}_h \in \mathbb{C}^h, | \, \widehat{y}_k^R \geq \mathbf{S}^R \right\} \text{ and } \left\{ \mathbf{z}_h \in \mathbb{C}^h, | \, \widehat{y}_k^R < \mathbf{S}^R \right\} \tag{6.32}$$

Similarly, the imaginary-part decision boundary $\mathbf{S}^I$ classifies the hidden layer output $\mathbf{z}_h \in \mathbb{C}^h$ into two decision regions

$$\left\{ \mathbf{z}_h \in \mathbb{C}^h, | \, \widehat{y}_k^I \geq \mathbf{S}^I \right\} \text{ and } \left\{ \mathbf{z}_h \in \mathbb{C}^h, | \, \widehat{y}_k^I < \mathbf{S}^I \right\} \tag{6.33}$$

The normal vector ($Q^R(\mathbf{h}^R, \mathbf{h}^I)$ to the real-part of the decision boundary ($\mathbf{S}^R$) and the normal vector ($Q^I(\mathbf{h}^R, \mathbf{h}^I)$ to the imaginary-part of the decision boundary ($\mathbf{S}^I$) are:

$$\begin{aligned} Q^R(\mathbf{z}_h^R, \mathbf{z}_h^I) &= \left(\frac{\partial \widehat{y}_k^R}{\partial z_h^{1R}} \cdots \frac{\partial \widehat{y}_k^R}{\partial z_h^{hR}} \frac{\partial \widehat{y}_k^R}{\partial z_h^{1I}} \cdots \frac{\partial \widehat{y}_k^R}{\partial z_h^{hI}} \right) \\ &= \left(v_{k1}^R \; \cdots \; v_{kh}^R \; - v_{k1}^I \; \cdots \; - v_{kh}^I \right) \end{aligned} \tag{6.34}$$

$$\begin{aligned} Q^I(\mathbf{z}_h^R, \mathbf{z}_h^I) &= \left(\frac{\partial \widehat{y}_k^I}{\partial z_h^{1R}} \cdots \frac{\partial \widehat{y}_k^I}{\partial z_h^{hR}} \frac{\partial \widehat{y}_k^I}{\partial z_h^{1I}} \cdots \frac{\partial \widehat{y}_k^I}{\partial z_h^{hI}} \right) \\ &= \left(v_{k1}^I \; \cdots \; v_{kh}^I \; v_{k1}^R \; \cdots \; v_{kh}^R \right) \end{aligned} \tag{6.35}$$

The decision boundaries ($\mathbf{S}^R$ and $\mathbf{S}^I$) of the kth output neuron are orthogonal iff the dot product of their normal vectors is zero.

$$Q^R(\mathbf{z}_h^R, \mathbf{z}_h^I).Q^I(\mathbf{z}_h^R, \mathbf{z}_h^I) = \begin{pmatrix} c_{k1}^R . v_{k1}^I + \cdots + v_{kh}^R . v_{kh}^I \\ -v_{k1}^I . v_{k1}^R + \cdots - v_{kh}^I . v_{kh}^R \end{pmatrix} \tag{6.36}$$

$$= 0. \tag{6.37}$$

From the above results, we can say that any output neuron in the CC-ELM has two decision boundaries with respect to the hidden neuron output and they are orthogonal to each other.

6.2.5 Case (ii): Orthogonality of Decision Boundaries in the Hidden Layer

The response of the jth neuron in the hidden layer can be written as

$$z_{h_j} = z^R_{h_j} + i z^I_{h_j} = sech(O_j),$$
$$\text{where } O_j = \boldsymbol{\sigma}^T_j\,(\mathbf{z} - \mathbf{c}_j), \quad j = 1, 2, \cdots, h \tag{6.38}$$
$$O_j = O^R_j + iO^I_j = (\boldsymbol{\sigma}^R_j + i\boldsymbol{\sigma}^I_j)^T\,\left[(\mathbf{z}^R - \mathbf{c}^R_j) + i\,(\mathbf{z}^I - \mathbf{c}^I_j)\right] \tag{6.39}$$
$$= \left[\boldsymbol{\sigma}^R_j\,(\mathbf{z}^R - \mathbf{c}^R_j) - \boldsymbol{\sigma}^I_j\,(\mathbf{z}^I - \mathbf{c}^I_j)\right] + i\left[\boldsymbol{\sigma}^R_j\,(\mathbf{z}^I - \mathbf{c}^I_j) + \boldsymbol{\sigma}^I_j\,(\mathbf{z}^R - \mathbf{c}^R_j)\right] \tag{6.40}$$

Using trigonometric and hyperbolic trigonometric definitions, the *sech* function can be written as:

$$sech(O^R_j + iO^I_j) = \frac{2\left(\cos(O^I_j\cosh(O^R_j) - i\sin(O^I_j)\sinh(O^R_j)\right)}{\cos(2O^I_j) + \cosh(2O^I_j)} \tag{6.41}$$

The two decision boundaries formed by the real and imaginary parts of the jth hidden neuron response with respect to the inputs are:

$$z^R_{h_j} = \frac{2\left(\cos(O^I_j\cosh(O^R_j)\right)}{\cos(2O^I_j) + \cosh(2O^I_j)} \rightarrow \mathbf{S}^R \tag{6.42}$$

$$z^I_{h_j} = \frac{2\left(-i\sin(O^I_j)\sinh(O^R_j)\right)}{\cos(2O^I_j) + \cosh(2O^I_j)} \rightarrow \mathbf{S}^I \tag{6.43}$$

Note that here the decision boundaries are shown with respect to the net input to the hidden neuron O_j, which is a linear function of actual input $\mathbf{z}$.

Hence, the complex-valued input $\mathbf{z} \in \mathbb{C}^m$ are classified into two decision regions with respect to the real and imaginary parts ($\mathbf{S}^R$ and $\mathbf{S}^I$)

$$\left\{\mathbf{z} \in \mathbb{C}^m, \mid z^R_{h_j} \geq \mathbf{S}^R\right\} \text{ and } \left\{\mathbf{z} \in \mathbb{C}^m, \mid z^R_{h_j} < \mathbf{S}^R\right\} \tag{6.44}$$
$$\left\{\mathbf{z} \in \mathbb{C}^m, \mid z^I_{h_j} \geq \mathbf{S}^I\right\} \text{ and } \left\{\mathbf{z} \in \mathbb{C}^m, \mid z^I_{h_j} < \mathbf{S}^I\right\} \tag{6.45}$$

The normal vectors to the decision boundaries formed by the hidden neuron (given by Eqs. (6.42) and (6.43)) with respect to the input ($\mathbf{z}$) are given by:

$$Q^R_h(\mathbf{z}^R, \mathbf{z}^I) = \left(\frac{\partial z^R_{h_j}}{\partial z^R_1} \cdots \frac{\partial z^R_{h_j}}{\partial z^R_k} \cdots \frac{\partial z^R_{h_j}}{\partial z^R_m}\,\frac{\partial z^R_{h_j}}{\partial z^I_1} \cdots \frac{\partial z^R_{h_j}}{\partial z^I_k} \cdots \frac{\partial z^R_{h_j}}{\partial z^I_m}\right) \tag{6.46}$$

$$Q^I_h(\mathbf{z}^R, \mathbf{z}^I) = \left(\frac{\partial z^I_{h_j}}{\partial z^R_1} \cdots \frac{\partial z^I_{h_j}}{\partial z^R_k} \cdots \frac{\partial z^I_{h_j}}{\partial z^R_m}\,\frac{\partial z^I_{h_j}}{\partial z^I_1} \cdots \frac{\partial z^I_{h_j}}{\partial z^I_k} \cdots \frac{\partial z^I_{h_j}}{\partial z^I_m}\right) \tag{6.47}$$

The decision boundaries are orthogonal to each other iff the dot product of these normal vectors is zero. The dot product of the normal vectors is given by:

$$Q_h^R(\mathbf{z}^R,\mathbf{z}^I).Q_h^I(\mathbf{z}^R,\mathbf{z}^I) = \sum_{k=1}^{m} \left(\frac{\partial z_{h_j}^R}{\partial z_k^R} \cdot \frac{\partial z_{h_j}^I}{\partial z_k^R} \right) + \left(\frac{\partial z_{h_j}^R}{\partial z_k^I} \cdot \frac{\partial z_{h_j}^I}{\partial z_k^I} \right) \tag{6.48}$$

$$\frac{\partial z_h^{jR}}{\partial z_k^R} = \frac{\partial z_h^{jR}}{\partial O_j^R}\frac{\partial O_j^R}{\partial z_k^R} + \frac{\partial z_h^{jR}}{\partial O_j^I}\frac{\partial O_j^I}{\partial z_k^R} = \frac{\partial z_h^{jR}}{\partial O_j^R}\sigma_{jk}^R + \frac{\partial z_h^{jR}}{\partial O_j^I}\sigma_{jk}^I \tag{6.49}$$

$$\frac{\partial z_h^{jI}}{\partial z_k^R} = \frac{\partial z_h^{jI}}{\partial O_j^R}\frac{\partial O_j^R}{\partial z_k^R} + \frac{\partial z_h^{jI}}{\partial O_j^I}\frac{\partial O_j^I}{\partial z_k^R} = \frac{\partial z_h^{jI}}{\partial O_j^R}\sigma_{jk}^R + \frac{\partial z_h^{jI}}{\partial O_j^I}\sigma_{jk}^I \tag{6.50}$$

$$\frac{\partial z_h^{jR}}{\partial z_k^I} = \frac{\partial z_h^{jR}}{\partial O_j^R}\frac{\partial O_j^R}{\partial z_k^I} + \frac{\partial z_h^{jR}}{\partial O_j^I}\frac{\partial O_j^I}{\partial z_k^I} = \frac{\partial z_h^{jR}}{\partial O_j^R}(-\sigma_{jk}^I) + \frac{\partial z_h^{jR}}{\partial O_j^I}\sigma_{jk}^R \tag{6.51}$$

$$\frac{\partial z_h^{jI}}{\partial z_k^I} = \frac{\partial z_h^{jI}}{\partial O_j^R}\frac{\partial O_j^R}{\partial z_k^I} + \frac{\partial z_h^{jI}}{\partial O_j^I}\frac{\partial O_j^I}{\partial z_k^I} = \frac{\partial z_h^{jI}}{\partial O_j^R}(-\sigma_{jk}^I) + \frac{\partial z_h^{jI}}{\partial O_j^I}\sigma_{jk}^R \tag{6.52}$$

Using the laws of differentiation, the derivative of the jth hidden layer response with respect to its net input

$$\frac{\partial h_j^R}{\partial O_j^R} = \frac{sinh(O_j^R)cos(3O_j^I) - sinh(3O_j^R)cos(O_j^I)}{\left(cos(2O_j^I) + cosh(2O_j^R)\right)^2}; \tag{6.53}$$

$$\frac{\partial h_j^I}{\partial O_j^I} = \frac{sinh(O_j^R)cos(3O_j^I) - sinh(3O_j^R)cos(O_j^I)}{\left(cos(2O_j^I) + cosh(2O_j^R)\right)^2} \tag{6.54}$$

$$\frac{\partial h_j^R}{\partial O_j^I} = \frac{cosh(O_j^R)sin(3O_j^I) - cosh(3O_j^R)sin(O_j^I)}{\left(cos(2O_j^I) + cosh(2O_j^R)\right)^2} \tag{6.55}$$

$$\frac{\partial h_j^I}{\partial O_j^R} = \frac{cosh(3O_j^R)sin(O_j^I) - cosh(O_j^R)sin(3O_j^I)}{\left(cos(2O_j^I) + cosh(2O_j^R)\right)^2} \tag{6.56}$$

From Eqs. (6.53)-(6.56), it can be observed that

$$\frac{\partial z_h^{j^R}}{\partial O_j^R} = \frac{\partial z_h^{j^I}}{\partial O_j^I} \tag{6.57}$$

$$\text{and } \frac{\partial z_h^{j^R}}{\partial O_j^I} = -\frac{\partial z_h^{j^I}}{\partial O_j^R} \tag{6.58}$$

Hence, it is evident that the *sech* activation function satisfies the Cauchy Riemann equations.

Substituting the Cauchy Riemann Equations in Eqs. (6.49)-(6.52), and obtaining the dot product of the normal vectors, we have,

$$Q_h^R(\mathbf{z}^R, \mathbf{z}^I).Q_h^I(\mathbf{z}^R, \mathbf{z}^I) = \sum_{k=1}^{m} \begin{bmatrix} -\frac{\partial z_h^{j^R}}{\partial O_j^R}\frac{\partial z_h^{j^R}}{\partial O_j^I}\left(\sigma_{jk}^R\right)^2 - \left(\frac{\partial z_h^{j^R}}{\partial O_j^I}\right)^2 \sigma_{jk}^R\sigma_{jk}^I \\ +\left(\frac{\partial z_h^{j^R}}{\partial O_j^R}\right)^2 \sigma_{jk}^R\sigma_{jk}^I + \frac{\partial z_h^{j^R}}{\partial O_j^R}\frac{\partial z_h^{j^R}}{\partial O_j^I}\left(\sigma_{jk}^I\right)^2 \\ -\frac{\partial z_h^{j^R}}{\partial O_j^R}\frac{\partial z_h^{j^R}}{\partial O_j^I}\left(\sigma_{jk}^I\right)^2 + \left(\frac{\partial z_h^{j^R}}{\partial O_j^I}\right)^2 \sigma_{jk}^R\sigma_{jk}^I \\ -\left(\frac{\partial z_h^{j^R}}{\partial O_j^R}\right)^2 \sigma_{jk}^R\sigma_{jk}^I + \frac{\partial z_h^{j^R}}{\partial O_j^R}\frac{\partial z_h^{j^R}}{\partial O_j^I}\left(\sigma_{jk}^R\right)^2 \end{bmatrix} \tag{6.59}$$

$$= 0. \tag{6.60}$$

Thus, the decision boundaries formed by the real and imaginary parts of the hidden layer output are orthogonal. It is also clear that this fact is also valid for any fully complex-valued activation function that satisfies the Cauchy Riemann equations.

Based on the above results, we state the following lemma:

Lemma 6.1. *The decision boundaries formed by the real and imaginary parts of an output/hidden neuron in a fully complex-valued network with any fully complex-valued activation function that satisfies the Cauchy Riemann conditions are orthogonal to each other.*

6.3 Summary

In this chapter, the various complex-valued classifiers available in the literature have been discussed and a fast learning circular complex-valued extreme learning machine classifier was described in detail. CC-ELM classifier uses a single hidden layer network with a non-linear input/hidden layer and a linear output layer. At the input layer, a unique nonlinear circular transformation is used as the activation function to make an one-to-one mapping of the real-valued input features to the Complex domain. At the hidden layer, the complex-valued input features are mapped on to a higher dimensional Complex plane using the '*sech*' activation function. In CC-ELM, the input parameters and the parameters of the hidden layer are chosen randomly and the output weights are calculated analytically, requiring lesser computational effort to perform classification tasks. The presence of orthogonal decision boundaries in CC-ELM are proved.

References

1. Nitta, T.: The Computational Power of Complex-Valued Neuron. In: Kaynak, O., Alpaydın, E., Oja, E., Xu, L. (eds.) ICANN 2003 and ICONIP 2003. LNCS, vol. 2714, pp. 993–1000. Springer, Heidelberg (2003)
2. Nitta, T.: Solving the xor problem and the detection of symmetry using a single complex-valued neuron. Neural Networks 16(8), 1101–1105 (2003)
3. Nitta, T.: Orthogonality of decision boundaries of complex-valued neural networks. Neural Computation 16(1), 73–97 (2004)
4. Aizenberg, I., Moraga, C.: Multilayer feedforward neural network based on multi-valued neurons (MLMVN) and a backpropagation learning algorithm. Soft Computing 11(2), 169–183 (2007)
5. Amin, M.F., Murase, K.: Single-layered complex-valued neural network for real-valued classification problems. Neurocomputing 72(4-6), 945–955 (2009)
6. Kim, T., Adali, T.: Fully complex multi-layer perceptron network for nonlinear signal processing. Journal of VLSI Signal Processing 32(1/2), 29–43 (2002)
7. Zhang, T.: Statistical behavior and consistency of classification methods based on convex risk minimization. Ann. Stat. 32(1), 56–85 (2003)
8. Suresh, S., Sundararajan, N., Saratchandran, P.: Risk-sensitive loss functions for sparse multi-category classification problems. Information Sciences 178(12), 2621–2638 (2008)
9. Savitha, R., Suresh, S., Sundararajan, N.: A fully complex-valued radial basis function network and its learning algorithm. International Journal of Neural Systems 19(4), 253–267 (2009)
10. Ortega, J.M.: Matrix Theory. Plenum Press, New York (1986)
11. Huang, G.B., Li, M.B., Chen, L., Siew, C.K.: Incremental extreme learning machine with fully complex hidden nodes. Neurocomputing 71(4-6), 576–583 (2008)
12. Suresh, S., Omkar, S.N., Mani, V., Prakash, T.N.G.: Lift coefficient prediction at high angle of attack using recurrent neural network. Aerospace Science and Technology 7(8), 595–602 (2003)

Chapter 7
Performance Study on Real-valued Classification Problems

As mentioned in Chapter 5, the orthogonal decision boundaries of fully complex-valued neural networks help them to perform classification tasks efficiently. Therefore, in this chapter, we study the classification performance of FC-MLP and IC-MLP described in Chapter 2, FC-RBF and Mc-FCRBF explained in Chapter 3, FCRN and CC-ELM described in the chapters 5 and 6 respectively. First, the study is conducted on a set of benchmark real-valued classification problems from the UCI machine learning repository [1] and then, using a practical acoustic emission signal classification problem for health monitoring [2].

7.1 Descriptions of Real-valued Benchmark Classification Problems

We consider a set of real-valued benchmark problems (both multi-category/binary classification problems) from the UCI machine learning repository [1]. Based on a wide range of Imbalance Factors (I. F.) (as defined in [3]) of the data set, three multi-category and four binary data sets are chosen for this study. To recap, the imbalance factor is defined as

$$(\text{I. F.}) = 1 - \frac{C}{N} \min_{j=1,\cdots,C} N_j \tag{7.1}$$

where N_j is the total number of samples belonging to the class j.

The detailed description of these data sets including the number of classes, the number of input features, the number of samples in the training/testing and the imbalance factor are presented in Table 7.1. From the table, one can see that the problems chosen for this study have both balanced and unbalanced data sets and also that the imbalance factors of the data sets vary over a wide range.

S. Suresh et al.: Supervised Learning with Complex-valued Neural Networks, SCI 421, pp. 125–133.
springerlink.com

Table 7.1 Description of benchmark data sets selected from [1] for performance study

Type of data set	Problem	No. of features	No. of classes	No. of samples		I. F.	
				Training	Testing	Training	Testing
Multi-Category	Image Segmentation (IS)	19	7	210	2100	0	0
	Vehicle Classification (VC)	18	4	424	422	0.1	0.12
	Glass Identification (GI)	9	7	109	105	0.68	0.73
Binary	Liver Disorder	6	2	200	145	0.17	0.145
	PIMA Data	8	2	400	368	0.225	0.39
	Breast Cancer	9	2	300	383	0.26	0.33
	Ionosphere	34	2	100	251	0.28	0.283

7.2 Performance Study

First we present the performance study results on three real-valued multi-category benchmark classification problems. Next, we consider four binary benchmark classification problems.

7.2.1 Performance Measures

The classification/confusion matrix Q is used to obtain the statistical measures for both the class-level and global performance of the various classifiers. Class-level performance is measured by the percentage classification (η_j) which is defined as:

$$\eta_j = \frac{q_{jj}}{N_j} \times 100\% \tag{7.2}$$

where q_{jj} is the total number of correctly classified samples in the class c_j.

The global measures used in the evaluation are the average per-class classification accuracy (η_a) and the over-all classification accuracy (η_o) defined as:

$$\eta_a = \frac{1}{C}\sum_{j=1}^{C} \eta_j$$

$$\eta_o = \frac{\sum_{j=1}^{C} q_{jj}}{\sum_{j=1}^{C} N_j} \times 100\% \tag{7.3}$$

The performance of the classifiers are compared using these class-level and global performance measures.

7.2.2 *Multi-category Real-valued Classification Problems*

As the complex-valued networks are shown to have better computational power than the real-valued networks [4], the classification performance of the complex-valued learning algorithms are compared against well-known real-valued classifiers, available in the literature for these problems. The real-valued classifiers used for comparison are the Support Vector Machines (SVM) [5], the minimal resource allocation network (MRAN) [6], the growing and pruning radial basis function network (GAP-RBFN) [7], the online sequential extreme learning machine (OS-ELM) [8], the real coded genetic algorithm based extreme learning machine [9], the Sequential Multi-Category Radial Basis Function (SMC-RBF) [10] and the Self-adaptive Resource Allocation Network (SRAN) [11]. The "*asinh*" and "*atan*" activation functions are observed to be better than the other ETF's and they are chosen as activation functions in the hidden layer for the FC-MLP and IC-MLP. The results of the RCGA-ELM is reproduced from [9], while those of the other real-valued classifiers are reproduced from [10]. The classification results for the PE-CVNN are reproduced from [12], while the results of the MLMVN are generated using the software simulator available in the author's web site [1].

Table 7.2 presents the overall and average testing efficiencies of the various classifiers on the three multi-category benchmark classification problems chosen for this study. In this study, the complex-valued input features (z) for FC-MLP, IC-MLP, FC-RBF and Mc-FCRBF are obtained by phase encoding the real-valued input features (x) in $[0,\pi]$ [13] using the transformation:

$$z = exp(i\phi) = cos\,\phi + isin\,\phi, \text{ where } \phi = \frac{\pi(x-a)}{b-a};\ a,b \in \mathbb{R} \text{ and } x \in [a,b]. \quad (7.4)$$

The input features for CC-ELM and FCRN classifiers are obtained by using the circular transformation defined in Eq. (6.19).

From the table, it is clear that the complex-valued classifiers outperform all the existing real-valued classifiers. The superior performance of the complex-valued classifiers can be attributed to their orthogonal decision boundaries. The higher performance of the complex-valued classifiers is very obvious in the glass identification problem which has a highly unbalanced data set.

Following observations emerge from the Table 7.2:

- FC-MLP, IC-MLP, FC-RBF, Mc-FCRBF, FCRN and CC-ELM classifiers outperform other complex-valued classifiers: MLMVN and PE-CVNN The performances of MLMVN [14] and PE-CVNN [13] may be limited because of the following factors:

[1] http://www.eagle.tamut.edu/faculty/igor/Downloads.htm

- The activation functions used in the PE-CVNN are similar to those used in the split complex-valued neural networks. Therefore, the correlation between the real and imaginary parts of the error are not considered in the network parameter update and the gradients are not fully complex-valued [15]. The limitations of using the split complex-valued activation functions have been discussed in detail in section 2.1.1.
- The complex-valued Multi Layer Multi Valued Network (MLMVN) that employs the multi-valued neurons uses a derivative free global error correcting learning rule to update the network parameters [14]. In MLMVN, the normalized real-valued input features (x) are mapped on to a full unit circle using $exp(i2\pi x)$ and the class labels are encoded by the roots of unity in the Complex plane. However, as the input features are mapped on to a full unit circle, this mapping results in the same complex-valued features for the real-valued features with values 0 and 1 (transformation is not unique). In addition, the multi-valued neurons map the complex-valued inputs to C discrete outputs on the unit circle. As number of classes (C) increases, the areas of the sectors per class within the unit circle decreases which results in a higher misclassification rate.

- Comparing the performances of FC-MLP and IC-MLP classifiers, IC-MLP classifier outperforms FC-MLP classifier in all the three benchmark problems considered. While the "*atan*" activation function resulted in a better classification of the IS data set, the "*asinh*" activation function outperforms the "*atan* " activation function in the classification of the unbalanced VC and GI data sets.
- FC-RBF classifier performs better than FC-MLP and IC-MLP classifiers in all the three problems. This is because the $sech(.)$ function used in FC-RBF has a magnitude response that is similar to that of the Gaussian function and has similar localization properties of the Gaussian activation function. This aids in improving the classification ability of FC-RBF classifier compared to that of the FC-MLP/IC-MLP classifiers.
- Mc-FCRBF classifier performs better than FC-RBF classifier along with a reduced computational effort. The self-regulatory system chose 155 of the total 210 samples for classification of the image segmentation problem which has a well-balanced data set. The self-regulatory system selects 358 of the total 422 samples and 272 of the total 336 samples to train Mc-FCRBF classifier for the vehicle classification and glass identification problems, respectively.
- CC-ELM classifier outperforms all the real-valued/complex-valued classifiers used in this study. It also requires the lowest computational effort in all the three real-valued benchmark classification problems. This can be attributed to the presence of the circular transformation that transforms the real-valued input features to all the four quadrants of the Complex domain uniquely and the learning algorithm that finds the optimum solution to the set of linear equations formed at the output layer. The best performance of CC-ELM can be distinctively seen in the glass identification problem which has a highly unbalanced data set.

- The performance of FCRN is slightly better than CC-ELM and is much better than the other classifiers. This can be attributed to the fact that FCRN uses the logarithmic error function, while the other classifiers use the mean squared error function.

7.2.3 Binary Real-valued Classification Problems

Next, we present the results of the binary benchmark classification problems listed in Table 7.1. Since it has been observed from the study on multi-category benchmark problems that FC-RBF and Mc-FCRBF outperform the FC-MLP and IC-MLP classifiers, we only compare the classification performances of FC-RBF and Mc-FCRBF classifiers in comparison with other real-valued classifiers. Performance results of SVM, ELM, SRAN, FC-RBF, Mc-FCRBF, FCRN and CC-ELM classifiers are presented in Table 7.3. From the results, one can see that the complex-valued classifiers, FC-RBF, Mc-FCRBF and CC-ELM classifiers outperformed the real-valued classifiers (SVM, ELM and SRAN) available in the literature. Among the complex-valued classifiers, CC-ELM and FCRN classifiers perform better than the other complex-valued classifiers considered in the study with the lower computational effort.

7.3 Performance Study Using a Real-world Acoustic Emission Classification Problem

Acoustic emission signals are the electrical versions of the stress or pressure waves produced by sensitive transducers. These waves are produced due to the transient energy release caused by the irreversible deformation processes in the material [2]. Different sources of acoustic emission exist and these sources can be characterized by the acoustic signals. The classification of acoustic emission signals based on their sources is a very difficult problem, especially in the real world where ambient noise and pseudo acoustic emission signals exist. Even in a noise free environment, superficial similarities exist between the acoustic emission signals produced by different sources making the classification task cumbersome.

In the study conducted in [2], noise free burst type acoustic emission signal from a metal surface is assumed. The data set presented in [2] uses 5 input features to classify the acoustic signals to one of the 4 sources, i.e., the pencil sourcc, the pulse source, the spark source and the noise. A training data set with 62 samples and testing data set with 137 samples are used for the acoustic emission signal classification problem. For details of the input features and the experimental set up used in the data collection, refer to [2].

Table 7.4 presents the performance results of the complex-valued FC-RBF, Mc-FCRBF, FCRN and CC-ELM classifiers in comparison to the best results available in the literature for the acoustic emission signal classification problem, viz., the Fuzzy K-means clustering algorithm [2], ant colony optimization algorithm [16]

Table 7.2 Benchmark classification problems: Performance comparison of the SR-FC-RBF classifier with other classifiers

Problem	Domain	Classifier	h	Time (sec.)	Testing η_o	Testing η_a
IS	Real	SVM	96	721	90.62	90.62
		MRAN	76	783	86.52	86.52
		GAP-RBFN	83	365	87.19	87.19
		OS-ELM	100	21	90.67	90.67
		RCGA-ELM	50	-	91	91
		SMC-RBF	43	142	91	91
		SRAN	47	22	92.3	92.3
	Complex	PE-CVNN	-	-	93.2[2]	-
		MLMVN	80	1384	83	-
		FC-MLP(*asinh*)	80	374	91.57	91.57
		FC-MLP(*atan*)	75	359	90.48	90.48
		IC-MLP(*asinh*)	80	390	91.81	91.81
		IC-MLP(*atan*)	80	385	92.81	92.81
		FC-RBF	38	421	92.33	92.33
		Mc-FCRBF	36	362	92.9	92.9
		FCRN	70	0.4	93.3	93.3
		CC-ELM	60	0.03	93.2	93.2
VC	Real	SVM	234	550	68.72	67.99
		MRAN	100	520	59.94	59.83
		GAP-RBFN	81	452	59.24	58.23
		OS-ELM	300	36	68.95	67.56
		SMC-RBF	75	120	74.18	73.52
		SRAN	113	55	75.12	76.86
	Complex	PE-CVNN	-	-	78.7[3]	-
		MLMVN	90	1396	78	77.25
		FC-MLP(*asinh*)	75	530	76.07	77.49
		FC-MLP (*atan*)	70	462	73.22	73.83
		IC-MLP(*asinh*)	75	612	79.62	80.38
		IC-MLP (*atan*)	70	574	74.17	74.26
		FC-RBF	70	678	77.01	77.46
		Mc-FCRBF	70	638	77.72	77.58
		FCRN	90	0.8	82.62	82.46
		CC-ELM	85	0.1084	82.23	82.52
GI	Real	SVM	102	320	64.23	60.01
		MRAN	51	520	63.81	70.24
		GAP-RBFN	75	410	58.29	72.41
		OS-ELM	60	15	67.62	70.12
		SMC-RBF	58	97	78.09	77.96
		SRAN	59	28	86.21	80.95
	Complex	PE-CVNN	-	-	65.5[b]	-
		MLMVN	85	1421	73.24	66.83
		FC-MLP(*asinh*)	70	338	80.95	79.60
		FC-MLP(*atan*)	70	346	80	79.09
		IC-MLP(*asinh*)	80	390	82.86	80.55
		IC-MLP(*atan*)	70	356	81.90	82.97
		FC-RBF	90	452	83.76	80.95
		Mc-FCRBF	85	364	83.91	80
		FCRN	90	0.25	94.5	88.3
		CC-ELM	100	0.08	94.44	84.52

Table 7.3 Performance comparison on benchmark binary classification problems

Problem	Classifier Domain	Classifier	h	Training Time (s)	Testing Efficiency (η_o)
Breast cancer	Real-valued	SVM	190	0.1118	94.20
		ELM	65	0.1442	96.28
		SRAN	7	0.17	96.87
	Complex-valued	FC-RBF	10	158.3	97.12
		Mc-FCRBF	10	125	97.4
		FCRN	15	0.16	97.4
		CC-ELM	15	0.0811	97.39
Iono-sphere	Real-valued	SVM	30	0.0218	90.18
		ELM	25	0.0396	88.78
		SRAN	21	3.7	90.84
	Complex-valued	FC-RBF	10	186.2	89.48
		Mc-FCRBF	10	152	90
		FCRN	15	0.0624	92.03
		CC-ELM	15	0.0312	92.43
Liver disorders	Real-valued	SVM	158	0.0972	68.24
		ELM	132	0.1685	71.79
		SRAN	91	3.38	66.9
	Complex-valued	FC-RBF	20	133	74.6
		Mc-FCRBF	20	112	76.6
		FCRN	10	0.05	75.86
		CC-ELM	10	0.059	75.5
PIMA data	Real-valued	SVM	209	0.205	76.43
		ELM	218	0.2942	76.54
		SRAN	97	12.24	78.53
	Complex-valued	FC-RBF	20	130.3	78.53
		Mc-FCRBF	20	103	79.89
		FCRN	15	0.125	80.71
		CC-ELM	20	0.073	81.25

and genetic programming [17]. The results show that the complex-valued classifiers outperform the real-valued classifiers considered in this study. It can also be seen that CC-ELM and FCRN classifiers required only 10 neurons to achieve an over-all testing efficiency of 99.27%, which is about 6% better than the best results reported in the literature for this problem. Thus, CC-ELM and FCRN perform an efficient classification of the acoustic emission signals using a compact network.

Table 7.4 Performance comparison results for the acoustic emission problem

Classifier domain	Classifier	Testing	
		η_o	η_{av}
Real-valued	Fuzzy C-Means Clustering		93.34
Complex-valued	FC-RBF	96.35	95.2
	Mc-FCRBF	98.54	97.83
	FCRN	99.27	98.91
	CC-ELM	99.27	99.17

7.4 Summary

In this chapter, we studied the decision making ability of FC-MLP, IC-MLP, FC-RBF, Mc-FCRBF, FCRN and CC-ELM learning algorithms in comparison to other complex-valued classifiers, MLMVN and PE-CVNN. The study was performed using a set of multi-category and binary benchmark classification data sets from the UCI machine learning repository and a practical acoustic emission classification problem. Performance results show that the performance of the complex-valued classifiers are better than the real-valued classifiers available in the literature. The orthogonal decision boundaries of the complex-valued classifiers help them to outperform the real-valued classifiers. However, the performance of the complex-valued classifiers are affected due to the transformation used to convert the real-valued input features to the Complex domain, the activation function used at the hidden layer, and the nature of the learning algorithm. It was also observed that the circular transformation, which maps the real-valued input features to the Complex domain uniquely, is better than the phase encoded transformation and the bilinear transformation.

References

1. Blake, C., Merz, C.: UCI repository of machine learning databases. Department of Information and Computer Sciences. University of California, Irvine (1998), http://archive.ics.uci.edu/ml/
2. Omkar, S.N., Suresh, S., Raghavendra, T.R., Mani, V.: Acoustic emission signal classification using fuzzy C-means clustering. In: Proc. of the ICONIP 2002, 9th International Conference on Neural Information Processing, vol. 4, pp. 1827–1831 (2002)

3. Suresh, S., Sundararajan, N., Saratchandran, P.: Risk-sensitive loss functions for sparse multi-category classification problems. Information Sciences 178(12), 2621–2638 (2008)
4. Nitta, T.: The Computational Power of Complex-Valued Neuron. In: Kaynak, O., Alpaydın, E., Oja, E., Xu, L. (eds.) ICANN 2003 and ICONIP 2003. LNCS, vol. 2714, pp. 993–1000. Springer, Heidelberg (2003)
5. Cristianini, N., Taylor, J.S.: An Introduction to Support Vector Machines. Cambridge University Press, Cambridge (2000)
6. Yingwei, L., Sundararajan, N., Saratchandran, P.: Performance evaluation of a sequential minimal radial basis function (RBF) neural network learning algorithm. IEEE Transactions on Neural Networks 9(2), 308–318 (1998)
7. Huang, G.B., Saratchandran, P., Sundararajan, N.: An efficient sequential learning algorithm for growing and pruning RBF (GAP-RBF) networks. IEEE Transactions on Systems, Man, and Cybernetics - Part B: Cybernetics 34(6), 2284–2292 (2004)
8. Liang, N.-Y., Huang, G.B., Saratchandran, P., Sundararajan, N.: A fast and accurate online sequential learning algorithm for feedforward networks. IEEE Transactions on Neural Networks 17(6), 1411–1423 (2006)
9. Suresh, S., Venkatesh Babu, R., Kim, H.J.: No-reference image quality assessment using modified extreme learning machine classifier. Applied Soft Computing 9(2), 541–552 (2009)
10. Suresh, S., Sundararajan, N., Saratchandran, P.: A sequential multi-category classifier using radial basis function networks. Neurocomputing 71(7-9), 1345–1358 (2008)
11. Suresh, S., Dong, K., Kim, H.J.: A sequential learning algorithm for self-adaptive resource allocation network classifier. Neurocomputing 73(16–18), 3012–3019 (2010)
12. Amin, M.F., Islam, M.M., Murase, K.: Ensemble of single-layered complex-valued neural networks for classification tasks. Neurocomputing 72(10-12), 2227–2234 (2009)
13. Amin, M.F., Murase, K.: Single-layered complex-valued neural network for real-valued classification problems. Neurocomputing 72(4-6), 945–955 (2009)
14. Aizenberg, I., Moraga, C.: Multilayer feedforward neural network based on multi-valued neurons (MLMVN) and a backpropagation learning algorithm. Soft Computing 11(2), 169–183 (2007)
15. Kim, T., Adali, T.: Fully complex multi-layer perceptron network for nonlinear signal processing. Journal of VLSI Signal Processing 32(1/2), 29–43 (2002)
16. Omkar, S.N., Karanth, U.R.: Rule extraction for classification of acoustic emission signals using ant colony optimisation. Engineering Applications of Artificial Intelligence 21(8), 1381–1388 (2008)
17. Suresh, S., Omkar, S.N., Mani, V., Menaka, C.: Classification of acoustic emission signal using Genetic Programming. Journal of Aerospace Science and Technology 56(1), 26–41 (2004)

Chapter 8
Complex-valued Self-regulatory Resource Allocation Network (CSRAN)

All the algorithms described in the chapters 2, 3, 4 and 6, viz., FC-MLP, IC-MLP, CRBF, FC-RBF, Mc-FCRBF, FCRN and CC-ELM are batch learning algorithms. These algorithms require the complete training data set and the network structure has to be fixed a priori. However

- In most practical applications, either the complete training data set may not be available a priori, or the training data set may be too large. Some of the practical problems like the cancer classifications, human action recognition human behavior prediction etc allow temporal changes to the task being learnt. Hence, these batch learning algorithms may not be usable and one have to look for online/sequential learning algorithms for these applications.
- Another important issue in a neural network learning scheme is the selection of network architecture. One has to find a minimal architecture that accurately fits the true function described by the training data. A large network may accurately fit the training data, but may have poor generalization performance due to overtraining and also resulting in an increased computation time. On the other hand, a smaller network requires lesser computational effort but might not be able to approximate the true function accurately. The complexity is further increased when the training data set is not available a priori. Hence, sequential learning algorithms that evolve the network architecture by themselves during the learning process are preferred over batch learning schemes.

These issues in the batch learning algorithms have motivated the development of sequential learning schemes for neural networks. In a sequential/online learning mode, samples are trained one-by-one and are discarded after they are learnt. Also, the network structure evolves during the learning process by adding and deleting neurons as it acquires information from the training data set.

Some of the complex-valued sequential learning algorithms available in the literature include the Complex-valued Minimal Resource Allocation Network (CM-RAN) [1], and the Complex-valued Growing And Pruning Network (CGAP-RBF) [2]. They are the direct extensions of the real-valued Minimal Resource Allocation Network (MRAN) [3], and the Growing And Pruning Radial Basis Function

S. Suresh et al.: Supervised Learning with Complex-valued Neural Networks, SCI 421, pp. 135–168.
springerlink.com

network (GAP-RBF) [4] to the Complex domain, respectively. These algorithms add and prune neurons to the network until they achieve a parsimonious structure. Though these algorithms are complex-valued learning algorithms, Gaussian activation function (whose response is real-valued for complex-valued inputs) is employed at the hidden layer and they lack fully complex-valued activation functions. Besides, these algorithms use the real-valued extended Kalman filter for parameter updates during learning. In addition, these algorithms use the split complex-valued Real and Imaginary parts of the error to update the Real and Imaginary parts of the network parameters, respectively. Hence, the gradients used in these algorithms are not fully complex-valued. This results in a significant loss of complex-valued information during the backward computation, due to the loss of correlation between the real and imaginary parts of the signals, thereby affecting the magnitude and phase approximation performance of the network. Another important invalidated assumption that these algorithms rely on is that the inputs are uniformly distributed, which is not always true, especially in real world problems. Training a network with similar samples might result in overtraining of the training dataset, affecting the generalization ability of the network.

Generalization performance of the complex-valued neural networks are influenced by the following two major factors:

- In general, the learning algorithms are based on the assumption that the training data is uniformly distributed in the input space which is not always true, especially, in practical problems. Learning similar samples repeatedly may result in overtraining, thereby, affecting the generalization capability with an increased computational effort. Hence, there is a need to develop a learning algorithm that is capable of deciding *when-to-learn*, *what-to-learn* and *how-to-learn* the samples in the training data set.
- In all the above complex-valued networks, the mean square magnitude error has been used as a minimization criterion for the parameter updates in the learning algorithms. In the context of complex-valued signals, though the minimization of magnitude error implicitly minimizes the phase error to some extent, it will be advantageous to use the phase error explicitly in the minimization criterion for a better phase approximation [5].

When the complete training data set is not available for batch learning and the data arrives sequentially, the above mentioned issues complicate the problem further. To address these issues, in this chapter, we present a sequential learning algorithm for complex-valued neural network with a self-regulating mechanism. The self-regulating scheme controls the learning process and uses the magnitude/phase errors explicitly for better performances in both complex-valued function approximation and classification problems.

To overcome these issues in the complex-valued sequential learning algorithms, in this chapter, we develop a "Complex-valued Self-regulating Resource Allocation Network (CSRAN)" and its sequential learning algorithm. CSRAN algorithm handles the training samples one-by-one and discards them after learning. The basic building block of CSRAN is a fully complex-valued radial basis function network with a '*sech*' activation function. The network parameters are updated using a fully Complex-valued Extended Kalman Filter (C-EKF) scheme presented in [6]. CSRAN starts with no hidden neuron and adds/prunes neurons based on criteria involving both the magnitude and phase errors. CSRAN employs a self-regulating scheme to control the learning process. The self-regulating scheme in CSRAN decides *what-to-learn*, *when-to-learn* and *how-to-learn* based on the information present in the training samples. When the training samples arrive one-by-one, the self-regulating scheme performs one of the following three actions: a) *Sample deletion* without learning b) *Sample learning* (either add/prune the hidden neuron or update the network parameters) and c) *Sample reserve* (shift the sample for future use). Thus, CSRAN algorithm efficiently addresses three key issues namely, *when-to-learn*, *what-to-learn* and *how-to-learn*.

Performance of CSRAN algorithm is evaluated using a synthetic complex-valued function approximation problem and two real-world problems. First, we highlight the key features of CSRAN algorithm in detail using a synthetic Complex function approximation problem. Next, we present the performance comparison based on a complex-valued Quadrature Amplitude Modulation (QAM) non-minimum, non-linear phase equalization problem [7] and an adaptive beam-forming problem [8]. In these problems, the performance of CSRAN is compared with existing complex valued neural networks. The results indicate that CSRAN algorithm provides better magnitude and (especially) phase approximation and generalization with fewer samples for training and a compact network structure.

As mentioned earlier, the better computational ability and the orthogonal decision boundaries of the complex-valued neural networks render them with an exceptional ability to perform classification tasks. Hence, to evaluate the classification ability of CSRAN, we study the performance of CSRAN in solving two real-valued multi-category benchmark classification problems with unbalanced data sets. To solve the real-valued classification problem, the complex-valued input features are obtained by phase encoding the real-valued input features as explained in [9]. Here, we compare the performance of CSRAN with existing complex-valued and best-performing real-valued neural classifiers.

In this chapter, we first review the existing complex-valued sequential learning algorithms , viz., the complex-valued minimal resource allocation network (CM-RAN) [1], the complex-valued growing and pruning Network (CGAP-RBF) [2] and the incremental (fully complex-valued) extreme learning machines (I-ELM) [10], highlight their disadvantages and then explain the CSRAN learning algorithm in detail.

8.1 A Brief Review of Existing Complex-valued Sequential Learning Algorithms

In sequential learning algorithms, the samples in the training dataset are presented one-by-one and only once. Once presented and learnt, the samples are deleted from the training dataset. The size of the network evolves, as learning continues.

Neurons are added and deleted from the hidden layer, until the network achieves a compact network structure. Several sequential learning algorithms are available for real-valued networks. Some of these real-valued sequential learning algorithms have been extended to the Complex domain. Complex-valued minimal resource allocation network (CMRAN) [1], complex-valued growing and pruning (CGAP-RBF) [2] and the incremental fully complex-valued extreme learning machines (I-ELM (complex-valued)) [10] algorithm are a few sequential learning algorithms for the complex-valued RBF (CRBF) network, which are direct extension of the MRAN algorithm [3], GAP-RBF algorithm [4] and the Incremental Extreme Learning Machine (I-ELM) [11] for the real-valued RBF network, respectively. In this section, we briefly present these algorithms, first.

8.2 Complex-valued Minimal Resource Allocation Network (CMRAN)

The basic block for the CMRAN [1] sequential learning algorithm is the complex-valued RBF network. The structure of the CRBF network is shown earlier in Fig. 8.1. Let the training sample set of the function to be approximated be represented by $\{(\mathbf{z}^1, \mathbf{y}^1), (\mathbf{z}^2, \mathbf{y}^2), \cdots, (\mathbf{z}^t, \mathbf{y}^t), \cdots\}$, where $\mathbf{z}^t \in \mathbb{C}^m$ and $\mathbf{y}^t \in \mathbb{C}^n$. For the CMRAN algorithm, the neurons in the hidden layer employ the Gaussian activation function as the basis function, while the neurons in the output layer are linear. The response of neurons in the hidden layer is given by eq. (8.1).

$$y_h^j = \phi\left(\frac{\mathbf{z}^t - \mathbf{c}_j}{\sigma_j}\right) = \exp\left(-\frac{(\mathbf{z}^t - \mathbf{c}_j)^H(\mathbf{z}^t - \mathbf{c}_j)}{2\sigma_j^2}\right);\ j = 1, 2, \cdots h \tag{8.1}$$

and the output of the CMRAN is given by

$$\widehat{y}_k = \sum_{j=1}^{h} v_{kj} y_h^j;\ \text{where } k = 1,\ 2,\ \cdots n \tag{8.2}$$

where $\mathbf{z}^t \in \mathbb{C}^m$ are the set of complex-valued inputs at the input node, $\mathbf{c}_j \in \mathbb{C}^m$ is the complex-valued centers of the jth hidden neuron, $\sigma_j \in \mathbb{R}$ is the width of the Gaussian function for the jth neuron and v_{kj} is the complex-valued weight connecting the kth output neuron and the jth hidden neuron. In sequential learning algorithms, the network evolves during training, to finally achieve a compact structure, with neither too many nor too few neurons in the hidden layer. The CMRAN algorithm is one such complex-valued sequential learning algorithm that adds and prunes neurons, until it achieves a parsimonious network structure. As each sample is presented to the network, according to the CMRAN algorithm, the sample is either

- Used for adding a neuron to the network or
- Used to delete a neuron from the hidden layer of the network or
- Used in learning to update the network parameters

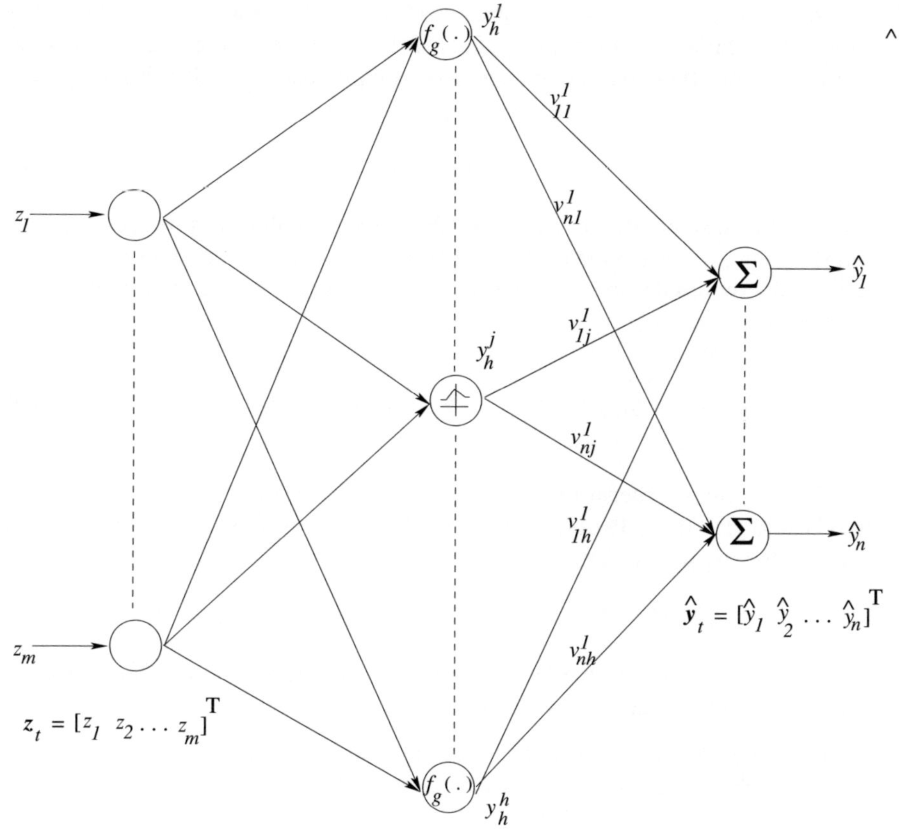

Fig. 8.1 Architecture of a CRBF Network

The CMRAN algorithm begins with zero hidden neurons. As each input-output training data $(\mathbf{z}^t, \mathbf{y}^t)$ is received, the error $\mathbf{e}^t = \mathbf{y}^t - \widehat{\mathbf{y}}^t$, where $\widehat{\mathbf{y}}^t = [\widehat{y}^1\ \widehat{y}^2\ \cdots\ \widehat{y}^n]^T$ is calculated. The algorithm adds or deletes hidden neurons, or adjusts the parameters of the existing network, according to the responses of the data received.

Growing condition: Let us consider $\mathbf{c}_{nr}$ is the center (of the hidden neuron) which is closest to $\mathbf{z}^t$ (the input of the data that was just received). If a sample presented to the network satisfies the following conditions, a new neuron is added to the network:

1. The squared distance between the sample and the nearest center must be greater than a threshold ε_n, as shown in eq. (8.3). This condition ensures that the new neuron that is added, is sufficiently far from all the existing neurons.

$$(\mathbf{z}^t - \mathbf{c}_{nr})^H (\mathbf{z}^t - \mathbf{c}_{nr}) > \varepsilon_n \tag{8.3}$$

where $\mathbf{c}_{nr}$ is the center of the neuron nearest to the sample considered and ε_n is appropriately chosen threshold.

2. The squared error of the sample presented must be greater than e_{min} (which is another threshold), as shown in eq. (8.4). This condition ensures that the existing neurons are insufficient to obtain a network output that meets the error specification.

$$\mathbf{e}^{t^H}\mathbf{e}^t > e_{min} \tag{8.4}$$

3. The root mean squared error over a window of M samples must be greater than a threshold e_{min1}, as shown in eq. (8.5). This condition checks whether the network has met the required sum squared error specification for the past M outputs of the network.

$$e_{rmst} = \sqrt{\frac{1}{M}\sum_{i=t-(M-1)}^{t} \mathbf{e}^{t^H}\mathbf{e}^t} > e_{min1} \tag{8.5}$$

Only when all these criteria (eqs. (8.3), (8.4), (8.5)) are met, will a new hidden neuron be added to the network. Each new hidden neuron added to the network will have the following parameters associated with it:

$$\mathbf{v}_{h+1} = \mathbf{e}^t; \quad \sigma_{h+1}^2 = \kappa(\mathbf{z}^t - \mathbf{c}_{nr})^H(\mathbf{z}^t - \mathbf{c}_{nr}); \quad \mathbf{c}_{h+1} = \mathbf{z}^t \tag{8.6}$$

where κ is the real valued overlap factor which determines the overlap of the responses of the hidden neurons in the input space.

Network learning: When an input to the network does not meet the criteria for adding a new hidden neuron, the network parameters

$$\mathbf{w}_{t-1} = [re(\mathbf{v}_1), im(\mathbf{v}_1), re(\mathbf{c}_1^T), im(\mathbf{c}_1^T), \sigma_1, \cdots re(\mathbf{v}_h), im(\mathbf{v}_h), re(\mathbf{c}_h^T), im(\mathbf{c}_h^T), \sigma_h]^T \tag{8.7}$$

are adapted using the extended Kalman filter as shown below:

$$\mathbf{w}^t = \mathbf{w}^{t-1} + \mathbf{klr}^t \times [Re(\mathbf{e}^t), Im(\mathbf{e}^t)]^T \tag{8.8}$$

where $Re(\mathbf{e}^t)$ denotes the Real part of the complex-valued error signal and $Im(\mathbf{e}^t)$ denotes the Imaginary part of the complex-valued error signal. In the EKF, a complex value is treated as a 2-dimensional real vector. $\mathbf{KLr}^t$ is the real-valued Kalman gain vector given by

$$\mathbf{KLr}^t = P^{t-1}\mathbf{a}^t[r + \mathbf{a}^{t^T}P^{t-1}\mathbf{a}^t]^{-1} \tag{8.9}$$

where r is the variance of the measurement noise, $\mathbf{a}^t$ is the gradient vector and has the following form

$$\left[\Delta\mathbf{v}_1, \Delta\mathbf{c}_1, \Delta\sigma_1, \cdots \Delta\mathbf{v}_h, \Delta\mathbf{c}_h, \Delta\sigma_h\right]^T$$

where $\Delta\mathbf{v}_1$, $\Delta\mathbf{c}_1$ and $\Delta\sigma_1$ are the gradient descent based updates for the three free parameters of the first neuron. Hence, the vector $\mathbf{a}^t$ is given by:

$$\begin{bmatrix} \Delta \mathbf{v}_1 \\ \Delta \mathbf{c}_1 \\ \Delta \sigma_1 \\ \vdots \\ \Delta \mathbf{v}_h \\ \Delta \mathbf{c}_h \\ \Delta \sigma_h \end{bmatrix} = \begin{bmatrix} I_{2\times 2}, \phi_1(\mathbf{z}^t) I_{2\times 2} \\ , \phi_1(\mathbf{z}^t)(2\beta_1/\sigma_1^2)[Re(\mathbf{z}^t - \mathbf{c}_1)^T, Im(\mathbf{z}^t - \mathbf{c}_1)^T], \\ \phi_1(\mathbf{z}^t)(2\beta_1/\sigma_1^3)(\mathbf{z}^t - \mathbf{c}_1)^H(\mathbf{z}^t - \mathbf{c}_1), \cdots, \\ \phi_h(\mathbf{z}^t) I_{2\times 2} \\ , \phi_h(\mathbf{z}^t)(2\beta_h/\sigma_h^2)[Re(\mathbf{z}^t - \mathbf{c}_h)^T, Im(\mathbf{z}^t - \mathbf{c}_h)^T], \\ \phi_h(\mathbf{z}^t)(2\beta_h/\sigma_h^3)(\mathbf{z}^t - \mathbf{c}_h)^H(\mathbf{z}^t - \mathbf{c}_h) \end{bmatrix}$$

Here, $\beta_1 = [Re(\mathbf{v}_1), Im(\mathbf{v}_1)]$, $\cdots$, $\beta_h = [Re(\mathbf{v}_h), Im(\mathbf{v}_h)]$. P^t is the error covariance matrix which is updated by

$$P^t = [I - \mathbf{KLr}^t \ \mathbf{a}^{t^T}]P^{t-1} + qI \tag{8.10}$$

where q is a scalar that determines the allowed random step in the direction of the gradient vector and I is an identity matrix. If the number of parameters to be adjusted is l then P^t is an $l \times l$ positive definite symmetric matrix. When a new hidden unit is allocated, the dimensionality of P^t increases to

$$\begin{pmatrix} P^{t-1} & 0 \\ 0 & p^0 I_0 \end{pmatrix}$$

where p_0 initializes the new rows and columns. p_0 is an estimate of the uncertainty in the initial values assigned to the parameters. The dimension of the identity matrix I_0 is equal to the number of new parameters introduced by the new hidden unit.

Pruning criterion: The algorithm also incorporates a pruning strategy which is used to prune hidden neurons that do not contribute significantly to the output of the network. This is done by observing the output of each of the hidden neurons for a pre-defined period and then removing the neuron that has not been producing a significant output for that period. Consider the output z_h^k of the kth hidden neuron:

$$y_h^k = \exp\left(-\frac{(\mathbf{z}^t - \mathbf{c}_k)^H(\mathbf{z}^t - \mathbf{c}_k)}{2\sigma_k^2}\right); \ k = 1, 2, \cdots h, \ \sigma_k \in \mathbb{R} \tag{8.11}$$

The response of a neuron k is small, if either σ_k of the above equation is small or if $(\mathbf{z}^t - \mathbf{c}_k)^H(\mathbf{z}^t - \mathbf{c}_k)$ is large. This would mean that the input is far away from the center of this hidden neuron. In any case, a small y_h^k means that its real part and imaginary part are both small. To reduce inconsistency caused by using the absolute values, both the real value and imaginary value of y_h^k are normalized with respect to the maximum (Real and Imaginary component of) output value among all the hidden neurons according to the following equation:

$$r_{hRe}^k = \frac{\| y_{hRe}^k \|}{\| y_{hRe}^{max} \|}; \ r_{hIm}^k = \frac{\| y_{hIm}^k \|}{\| y_{hIm}^{max} \|} \tag{8.12}$$

$\| y_{hRe}^{max} \|$ is the largest absolute real component of the hidden neuron output and $\| y_{hIm}^{max} \|$ is the largest absolute imaginary component of the hidden neuron output. If both r_{hRe}^k and r_{hIm}^k of a normalized output of y_h^k $(k < h)$ are less than a threshold, E_P, for S_w consecutive observations (where S_w is the width of the sliding window defining the number of samples to be considered), *i. e.*,

$$\text{If } r_{hRe}^k < E_P \quad r_{hIm}^k < E_P \text{ for } S_w \text{ consecutive observations} \tag{8.13}$$

then it indicates that the hidden neuron k makes insignificant contribution to the network output and can be removed from the network. The dimensions of the EKF are then adjusted to suit the reduced network. Thus, the algorithm ensures that the network has sufficient neurons in the hidden layer, and none of the neurons in the hidden layer is redundant. This gives a compact structure to the network.

The pseudocode for the CMRAN learning algorithm is presented in Pseudocode 2.

Pseudocode 2 Pseudo code for the CMRAN Algorithm.

Input: Present the training data one-by-one to the network
from datastream.
Output: Size of the network, parameters of the network.
START
Initialization: Assign the first sample as the first neuron($h=1$).
The parameters of the neuron are chosen as shown in eq. (8.6).
Start learning for samples $t=2,3,...$
DO
Compute the network output $\widehat{y}_t$.
Find the neuron nearest to the sample presented.
IF eq. (8.3) & eq. (8.4) & eq. (8.5) are satisfied **THEN**
Add a Neuron to the network (h=h+1).
Choose the parameters of the network as in eq. (8.6).
ELSE
Update the parameters of the network using EKF (eq. (8.8)).
Update the EKF parameters accordingly eq. (8.9), eq. (8.10)).
Pruning Criteria:
Calculate r_{hRe}^k and r_{hIm}^k according to eq. (8.12).
If $r_{hRe}^k < E_P$ and $r_{hIm}^k < E_P$ for S_w
consecutive observations, delete the neuron.
ENDIF
ENDDO
Stopping Strategy: Training stops when $t = N$
END

8.2.1 Drawbacks of the CMRAN Algorithm

However, the CMRAN algorithm suffers from the following drawbacks:

- The Gaussian function eq. (8.1) is used as the basis of the activation function of the network, used in these algorithms. This results in inaccurate phase approximations, despite the weights and centers being complex-valued, as shown in the simulation studies in [5, 12]. This results from the fact that the activation response of the network remains real-valued.
- The real-valued EKF is used for parameter updates. This uses the real and imaginary components of the complex-valued error and weights. This is not a true representation of the complex-valued error/weights and hence, approximation using the learning algorithm is not accurate.
- Proper selection of training dataset is an important task in the neural network training algorithms. In general, it is assumed that the training data is uniformly distributed in the input space with non-recurrent training samples. For most practical problems, it is difficult to satisfy this assumption. Hence, one needs to develop a learning algorithm which can select proper samples for learning. However, the CMRAN algorithm does not ensure that the samples are uniformly distributed in the input space. This affects the generalization ability of the network.
- The control parameters of the algorithm are problem dependent. A heuristic approach is used to choose the parameters of the algorithm.

8.3 Complex-valued Growing and Pruning RBF (CGAP-RBF) Networks

In this section, we present the CGAP-RBF learning algorithm [2] briefly. In the CGAP-RBF learning algorithm, the growing and pruning criteria for the hidden neurons is based on the concept of significance. When there is no growing or pruning, parameter adjustments using the EKF is done. The significance of a neuron is calculated as:

$$E_{sig}(k) = \frac{||\alpha_k||_2}{S(Z)^{1/2}} \left(\frac{\pi^{3/2}}{\sqrt(2)} \sigma_k \right)^{m/2} \tag{8.14}$$

where α_k is the weight connecting the kth hidden neuron to the output neurons, $S(Z)$ is the size of the input sampling space, assuming uniform distribution (in the input space) and $\sigma_k \in \mathbb{R}$ is the real-valued Gaussian width of the hidden neuron considered. For complete derivation of $E_{sig}(k)$, one must refer to [2]. During training, if $E_{sig}(k) < e_{goal}$, where e_{goal} is the target accuracy, then the average contribution made by the neuron h is less than the expected accuracy. This means that neuron k is insignificant and it can be removed. Similarly, for a newly added neuron $h+1$, if $E_{sig}(h+1) < e_{goal}$, it means that the newly added neuron makes insignificant contribution to the overall performance of the network (although it may make sense to that single-input observation). Hence, this neuron should not be added. An enhanced growing criterion of the CGAP-RBF algorithm is given by:

- ***Growing criterion:*** For a new training data observation $(\mathbf{z}^t, \mathbf{y}^t)$, if

$$\begin{aligned} \| \mathbf{z}^t - \mathbf{c}_{nr} \| &> \varepsilon_n \\ \text{and} E_{sig}(h+1) &> e_{goal} \end{aligned} \quad (8.15)$$

  where ε_n is a distance threshold for adding neurons, a new neuron $h+1$ should be added, and the parameters associated with the new hidden neuron are selected as follows

$$\begin{aligned} \mathbf{v}_{h+1} &= \mathbf{e}^t \\ \mathbf{c}_{h+1} &= \mathbf{z}^t \\ \sigma_{h+1} &= \sqrt{\kappa(\mathbf{z}^t - \mathbf{c}_{nr})^H(\mathbf{z}^t - \mathbf{c}_{nr})} \end{aligned} \quad (8.16)$$

  where $\mathbf{e}^t$ is the network error given by $\mathbf{y}^t - \widehat{\mathbf{y}}^t$, $\mathbf{c}_{nr}$ is the center of the nearest neuron to the sample considered and κ is the overlap factor that determines the overlap of the responses of the hidden neurons in the input space. It can be observed from eq. (8.15) that the selection of network parameters for a new hidden neuron added to is exactly similar to that of the CMRAN algorithm, as seen in eq. (8.6).
- ***Pruning criterion:*** If the significance of neuron k; $k < h$ is less than the accuracy e_{goal}, neuron k is insignificant and should be removed, otherwise, neuron k is significant and should be retained. The earlier mentioned condition implies that after learning each observation, the significance for all neurons should be computed and checked for possible pruning. This will be a computationally intensive task. However, only the nearest neuroncan possibly be insignificant and need to be checked for pruning, and there is no need to compute the significance for all neurons.
- ***Learning criterion:*** When no neuron is added or pruned from the network, in CGAP-RBF, only the parameters of the neuron nearest to the latest incoming observation will be adjusted using the EKF algorithm.

The CGAP-RBF algorithm is summarized in pseudocode 3.

It can be observed from this section that the CGAP-RBF algorithm is similar to the CMRAN learning algorithm. The differences between the CMRAN and the CGAP-RBF learning algorithms are:

- In the CMRAN, a new neuron is added only based on the distance between the sample and the neurons of the network. In the CGAP-RBF algorithm, in addition to the distance criterion, a new neuron will be added only if its significance is more than the chosen learning accuracy. Also, if during training, the significance for a neuron becomes less than the learning accuracy, then that neuron will be pruned.

- In the CMRAN, the free parameters of all the neurons are updated, if the add criterion is not satisfied. In the CGAP-RBF, if the input data does not require a new hidden neuron to be added, then the parameters of only the nearest neuron are adjusted, resulting in a reduction in the overall computations and thereby increasing the learning speed.

Pseudocode 3 Pseudo code for the CGAP-RBF Algorithm.

Input: Present the training data one-by-one to the network from datastream.
Output: Size of the network, parameters of the network.

```
START
  Initialization: Assign the first sample as the first neuron(h = 1).
      The parameters of the neuron are chosen as shown in eq. (8.16).
  Start learning for samples  t = 2, 3, ...
  DO
        Compute the network output ŷ_t.
        Find the neuron nearest to the sample presented.
     IF eq. (8.15) are satisfied THEN
        Add a Neuron to the network (h = h+1).
        Choose the parameters of the network as in eq. (8.16).
     ELSE
        Adjust the network parameters of the nearest neuron
        (v_nr, c_nr and σ_nr) only.
        Check the criterion for pruning the adjusted hidden
        neuron:
           If E_sig(nr) < e_goal
               Remove the hidden neuron(nr) that is nearest
               to the sample considered.
     ENDIF
  ENDDO
  Stopping Strategy: Training stops when t = N
END
```

However, the CGAP-RBF algorithm also suffers from all the issues stated in section 8.2 for the CMRAN, as it also uses the Gaussian activation function at the hidden layer, and the real-valued EKF for updating the free parameters of the network. Hence, approximation, especially phase approximation using the CGAP-RBF learning algorithm is also inaccurate, owing to these issues. Besides this, the calculation of significance of a hidden neuron assumes uniform distribution of samples in the input space, which is not always true, especially, in real-world problems. Hence, a fully complex-valued sequential learning algorithm with selective participation of samples in learning becomes essential. In this chapter, we present one such fully complex-valued sequential learning algorithm.

8.4 Incremental Fully Complex-valued Extreme Learning Machines (I-ELM)

Incremental fully complex-valued extreme learning machines [10] are direct extensions of the real-valued incremental extreme learning machines, available in literature. This algorithm assumes random initial weights and estimates the output weights analytically, similar to the extreme learning machines. In addition, the I-ELM algorithm adds and prunes neurons, until a desired accuracy is achieved, thus selecting the number of hidden neurons during learning. As the algorithm estimates the output weights analytically, the speed of learning is high. A brief summary of the I-ELM algorithm is presented in pseudocode 4.

Pseudocode 4 Pseudo code for the I-ELM Algorithm.

Input: Present the training data one-by-one
to the network from datastream,
maximum number of hidden nodes (h_max)
and expected learning accuracy (ε).

Output: Size of the network, parameters of the network.

START

Initialization: Assign $h = 0$ and error $E = Y$, $Y = [\mathbf{y}_1, \mathbf{y}_2, \ldots \mathbf{y}_N]$.

DO

while $h < h_{max}$ and $||E|| > \varepsilon$

a. Increase the number of hidden nodes $h = h+1$.

b. Assign the hidden node parameters $(\mathbf{c}_h, \sigma_h)$ for the new hidden node randomly.

c. Calculate the output weight ($\mathbf{v}_h$) for the new hidden neuron using

$$\mathbf{v}_h = \frac{E.z_h^{h*}}{z_h^h .z_h^{h*}}$$

d. Calculate the residual error E after adding the new hidden neuron \$h\$:

$$E = E - \mathbf{v}_h.z_h^h$$

ENDDO

END

8.5 Complex-valued Self-regulatory Resource Allocation Network Learning Algorithm (CSRAN)

In this section, we first describe CSRAN network architecture and then the principles behind the self-regulating learning scheme.

8.5.1 Network Architecture

The basic building block of CSRAN is a single hidden layer fully complex-valued radial basis function network as shown in Fig. 8.2. CSRAN is a sequential learning

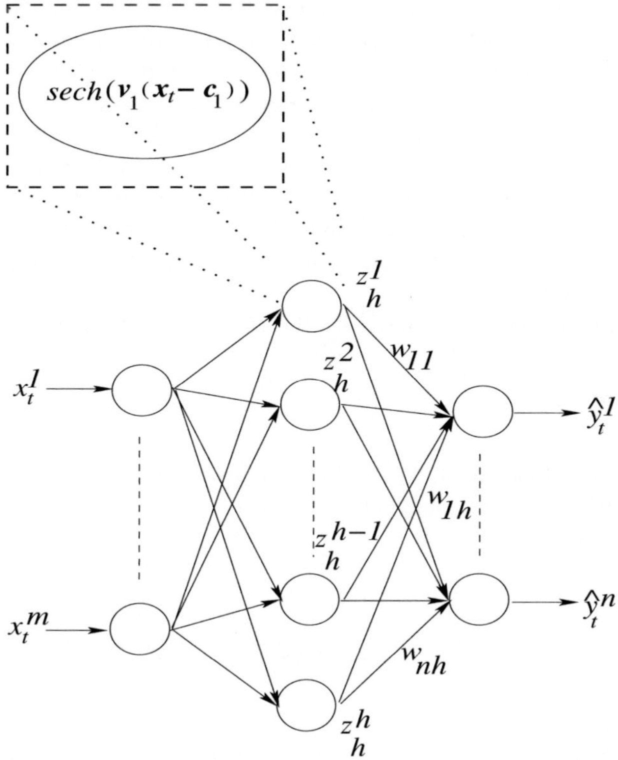

Fig. 8.2 Architecture of CSRAN

algorithm and starts with no hidden neuron and builds the necessary number of hidden neurons based on the information contained in the training samples. CSRAN has a self-regulating scheme which controls the learning process by proper selection of the training samples.

Let the sequence of training data drawn randomly be $\{(\mathbf{z}^1,\mathbf{y}^1), (\mathbf{z}^2,\mathbf{y}^2),\cdots, (\mathbf{z}^t,\mathbf{y}^t),\cdots\}$, where $\mathbf{z}^t = [z_1^t, \cdots, z_m^t]^T \in \mathbb{C}^m$ is an m-dimensional input vector and $\mathbf{y}^t = [y_1^t, \cdots, y_n^t]^T \in \mathbb{C}^n$ is its n-dimensional desired output.

Without loss of generality, we assume that after sequentially learning $t-1$ observations, CSRAN has built a network with h hidden neurons. For a given input, the predicted output is given by ($\widehat{\mathbf{y}}^t = [\widehat{y}_1^t, \cdots, \widehat{y}_n^t]^T \in \mathbb{C}^n$) of CSRAN with h hidden neurons is given by

$$\widehat{y}_j^t = \sum_{k=1}^{h} v_{jk} z_h^k; \; j = 1, \cdots, n \tag{8.17}$$

$$z_h^k = sech\left[\boldsymbol{\sigma}_k^T (\mathbf{z}^t - \mathbf{c}_k)\right]; \; k = 1, 2, \cdots h$$

where $v_{jk} \in \mathbb{C}$ is the complex-valued weight connecting the kth hidden neuron to the jth output neuron, $\boldsymbol{\sigma}_k = [\sigma_{k1}, \cdots, \sigma_{km}]^T \in \mathbb{C}^m$ is the complex-valued scaling factor of the kth hidden neuron, $\mathbf{c}_k = [c_{k1}, \cdots, c_{km}]^T \in \mathbb{C}^m$ is the center of the kth hidden neuron, the superscript T denotes the transpose operator and $sech(z) = 2/(e^z + e^{-z})$.

The magnitude response of the '$sech(\boldsymbol{\sigma}_k^T(\mathbf{z}^t - \mathbf{c}_k))$' activation function is similar to that of the Gaussian activation function. Also, the phase response of this function is close to zero when both the real and imaginary parts of the complex-valued signal $(\boldsymbol{\sigma}_k^T(\mathbf{z}^t - \mathbf{c}_k))$ are zero. Thus the activation function response is highest when the inputs are located closer to the center, making the function a good choice for a radial basis function.

The residual error ($\mathbf{e}^t$) of CSRAN for the current observation ($\mathbf{z}^t$, $\mathbf{y}^t$) is defined by

$$\mathbf{e}^t = \mathbf{y}^t - \widehat{\mathbf{y}}^t \tag{8.18}$$

From the above residual error, we estimate the instantaneous magnitude error (M_t^e) and the normalized absolute phase error (ϕ_t^e) as

$$M_t^e = \frac{1}{n}\sqrt{\mathbf{e}^{t^H}.\mathbf{e}^t} \tag{8.19}$$

$$\phi_t^e = \frac{1}{n\pi}\sum_{l=1}^{n}\left|arg(y_l^t) - arg(\widehat{y}_l^t)\right| \tag{8.20}$$

where the superscript H denotes the Complex Hermitian operator and $arg(.)$ is a function that returns the phase of a complex-valued number in $[-\pi, \pi]$, given by:

$$arg(z) = atan\left(\frac{imag(z)}{real(z)}\right) \tag{8.21}$$

8.5.2 Sequential Self-regulating Learning Scheme of CSRAN

In a sequential learning framework, the observation data/samples arrive one-by-one and one at a time. Most of the sequential learning algorithms (both in real-valued/complex-valued) learn all the training samples in the sequence as they are presented, whereas the proposed CSRAN algorithm regulates the sequential learning process by selecting appropriate samples for learning. The schematic diagram of the self-regulating scheme is shown in Fig. 8.3 and the basic working principles of the above scheme are explained in the following paragraphs.

Based on the instantaneous magnitude (M_t^e) and absolute phase error (ϕ_t^e) of each sample in the training sequence, the self-regulating scheme performs one of the following three actions:

Action (a) *Sample Deletion*: Samples are deleted without being used in the learning process.
Action (b) *Sample Learning*: Learning the sample which includes growing/pruning the hidden neurons or updating the existing network parameters.

Action (c) *Sample Reserve*: The samples are pushed to the rear end of the training sequence and can be used at a later stage.

The concepts behind these actions representing each block in Fig. 8.3 are described in detail below:

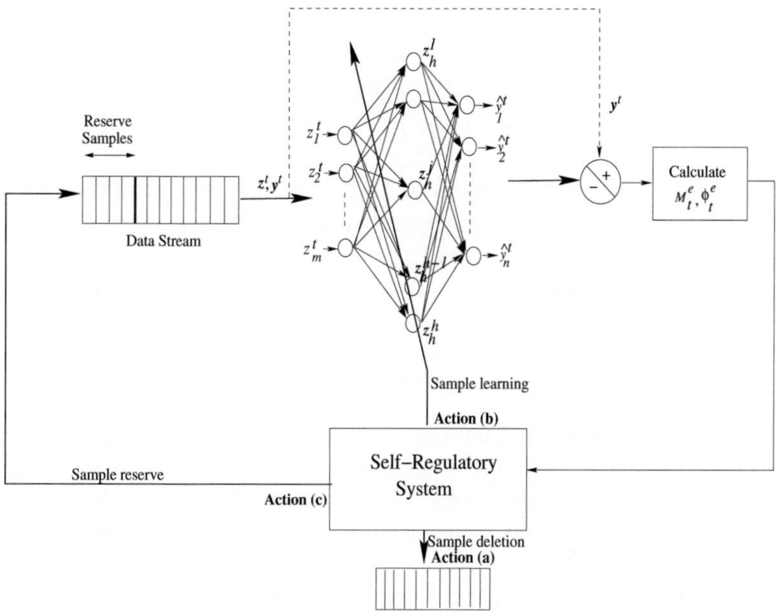

Fig. 8.3 Schematic Diagram of CSRAN and Its Self-regulating Learning Scheme

Action (a) Sample Deletion: When both the instantaneous magnitude error (M_t^e) (given in Eq. (8.19)) and the absolute phase error (ϕ_t^e) (given in Eq. (8.20)) of a sample are lesser than their fixed (deletion) thresholds, the self-regulating scheme deletes the sample without using it in the learning process. The sample deletion criterion is given by

$$M_t^e \le E_d^M \textbf{ AND } \phi_t^e \le E_d^\phi \tag{8.22}$$

where E_d^M is the sample deletion magnitude threshold and E_d^ϕ is the sample deletion phase threshold. The 'sample deletion' criterion removes similar samples from the training sequence. Hence, it avoids over-training and also reduces the computational effort.

Action (b) Sample Learning: In a self-regulating scheme, the learning process involves the allocation of new hidden neurons ('growing') or updating of network parameters ('update') and removing redundant neurons ('pruning').

Neuron Growing Criterion: As training samples arrive sequentially, some of the selected samples will be used to 'add' a new hidden neuron based on the following criterion

$$M_t^e \geq E_a^M \ \textbf{OR} \ \phi_t^e \geq E_a^\phi \tag{8.23}$$

where E_a^M is the neuron growing magnitude threshold and E_a^ϕ is the neuron growing phase threshold.

It should be pointed out here that the neuron growing thresholds (E_a^M, E_a^ϕ) are not kept constant. They are adaptively varied based on the current residual error as given below

$$\begin{aligned} &\text{IF} \ \ M_t^e \geq E_a^M \ \text{THEN} \ E_a^M := \delta E_a^M - (1-\delta) M_t^e \\ &\text{IF} \ \ \phi_t^e \geq E_a^\phi \ \text{THEN} \ E_a^\phi := \delta E_a^\phi - (1-\delta) \phi_t^e \end{aligned} \tag{8.24}$$

where the slope parameter (δ) controls the rate at which the neuron growing thresholds (E_a^M, E_a^ϕ) are regulated and hence influence the neuron growth. In general, the slope parameter is initialized close to 1.

When a new hidden neuron ($h+1$) is added to the network, the parameters associated with it are initialized as

$$\mathbf{v}_{h+1} = \mathbf{e}^t; \quad \mathbf{c}_{h+1} = \mathbf{z}^t; \quad \boldsymbol{\sigma}_{h+1} = \kappa(\mathbf{z}^t - \mathbf{c}_{nr}) \tag{8.25}$$

where nr is the nearest neuron, defined as that neuron with the smallest Euclidean distance from the current sample. The scaling factor κ determines the overlap between the samples in the input space. As κ increases, the overlap between the responses of the hidden neurons also increases. **Network Parameter Update Criterion:** If a new observation ($\mathbf{z}^t, \mathbf{y}^t$) arrives and the parameter update criterion is satisfied then the parameters of the network are updated using a C-EKF [6]. The parameter update criterion is given by

$$M_t^e \geq E_l^M \ \ \textbf{OR} \ \ \phi_t^e \geq E_l^\phi \tag{8.26}$$

where E_l^M is the parameter update magnitude threshold and E_l^ϕ is the parameter update phase threshold. The parameter update thresholds (E_l^M, E_l^ϕ) are also adapted based on the residual error of the current sample as given below

$$\begin{aligned} &\text{IF} \ \ M_t^e \geq E_l^M \ \text{THEN} \ E_l^M := \delta E_l^M - (1-\delta) M_t^e \\ &\text{IF} \ \ \phi_t^e \geq E_l^\phi \ \text{THEN} \ E_l^\phi := \delta E_l^\phi - (1-\delta) \phi_t^e \end{aligned} \tag{8.27}$$

where δ is a slope that controls the rate of self-adaptation of the parameter update magnitude and phase thresholds. Usually, δ is set close to 1.

The main advantage of the self-regulating thresholds is that it helps in selecting appropriate samples to add neuron or update the network parameters, i.e., CSRAN algorithm uses sample with higher error to either add a new hidden neuron or update the network parameters first and the remaining samples later for fine tuning the network parameters.

The network parameters ($\boldsymbol{\alpha}^t = [\mathbf{c}_1, \cdots, \mathbf{c}_h,$
$bm\sigma_1, \cdots, \boldsymbol{\sigma}_h, \mathbf{v}_1, \cdots, \mathbf{v}_h]^T$) are updated for the current sample (t) as

$$\boldsymbol{\alpha}^t = \boldsymbol{\alpha}^{t-1} + G^t\ \mathbf{e}^t \tag{8.28}$$

where $\mathbf{e}^t$ is the residual error and G^t is complex-valued Kalman gain matrix given by

$$G^t = P^{t-1}\mathbf{a}^t \left[R + \mathbf{a}^{t^H}\ P^{t-1}\ \mathbf{a}^t\right]^{-1} \tag{8.29}$$

where $\mathbf{a}^t$ is the complex-valued gradient vector, $R = r_0 I_{n\times n}$ is the variance of the measurement noise and P^t is the error covariance matrix.

The gradient vector ($\mathbf{a}^t$) (set of partial derivatives of output with respect to $\boldsymbol{\alpha}^t$) is defined as

$$\mathbf{a}^t = \begin{bmatrix} \frac{\partial \widehat{y}_1^t}{\partial c_{11}} \cdots \frac{\partial \widehat{y}_1^t}{\partial c_{hm}} & \frac{\partial \widehat{y}_1^t}{\partial \sigma_{11}} \cdots \frac{\partial \widehat{y}_1^t}{\partial \sigma_{hm}} & \frac{\partial \widehat{y}_1^t}{\partial v_{11}} \cdots \frac{\partial \widehat{y}_1^t}{\partial v_{nh}} \\ \vdots \quad\quad \vdots & \vdots \quad\quad \vdots & \vdots \quad\quad \vdots \\ \frac{\partial \widehat{y}_n^t}{\partial c_{11}} \cdots \frac{\partial \widehat{y}_n^t}{\partial c_{hm}} & \frac{\partial \widehat{y}_n^t}{\partial \sigma_{11}} \cdots \frac{\partial \widehat{y}_n^t}{\partial \sigma_{hm}} & \frac{\partial \widehat{y}_n^t}{\partial v_{11}} \cdots \frac{\partial \widehat{y}_n^t}{\partial v_{nh}} \end{bmatrix}^T$$

where,

$$\begin{aligned} \frac{\partial \widehat{y}_l^t}{\partial c_{jk}} &= \overline{v}_{lj}\ \overline{u}_h^j\ \overline{\boldsymbol{\sigma}}_{jk};\ u_h^j = \ -z_h^j.tanh(\boldsymbol{\sigma}_j^T(\mathbf{z}^t - \mathbf{c}_j)) \\ \frac{\partial \widehat{y}_l^t}{\partial \sigma_{jk}} &= \overline{v}_{lj}\ \overline{u}_h^j\ (\overline{\mathbf{z}_k^t - c_{jk}}) \\ \frac{\partial \widehat{y}_l^t}{\partial v_{lj}} &= \overline{z}_h^j;\ k = 1,\cdots,m; j = 1,\cdots,h; l = 1,\cdots,n \end{aligned} \tag{8.30}$$

$\overline{u}_h^j$ is the conjugate of the derivative of the jth hidden neuron output and $\overline{z}_h^j$ is the conjugate of the jth hidden neuron output. The error covariance matrix is updated as

$$P^t = \left[I - G^t\mathbf{a}^{t^H}\right]P^{t-1} + qI \tag{8.31}$$

where q is a process noise covariance usually set close to 0 and I is an identity matrix of dimension $h(2m+n) \times h(2m+n)$.

Neuron Pruning Criterion: Similar to CMRAN, the proposed CSRAN algorithm also uses the contribution of the hidden neuron to delete superfluous neurons. The contribution of the jth hidden neuron is defined as

$$r_k = \frac{z_h^k}{\max_j z_h^j} \tag{8.32}$$

If $\|r_k\| < E_p$ AND $arg(r_k) < E_p$ for N_w consecutive samples, then the kth neuron is superfluous and is removed from the network. Here, E_p is the neuron pruning threshold. If the neuron pruning threshold (E_p) is set at a lower value, then pruning seldom occurs and all the added neurons will remain in the network irrespective of their contribution to the network output. On the other hand, higher value of E_p

results in frequent pruning, resulting in oscillations and insufficient neurons to approximate the function.

When a neuron is added to the network, the error covariance matrix (P^t) is updated according to:

$$P^t = \begin{bmatrix} P^{t-1} & 0 \\ 0 & p^0 I_{(2m+n)\times(2m+n)} \end{bmatrix} \tag{8.33}$$

where p^0 is the estimated uncertainty of the initial parameters. On the other hand, when a neuron (say, k^{th} neuron) is removed from the network, the dimensionality of the error covariance matrix is reduced by removing the respective rows and columns of the P^t matrix, *i.e.* remove $(k-1)(2m+n)+1$ to $k(2m+n)$ rows and columns of the P^t matrix. The values of p_0, q and r_0 in the C-EKF are usually set close to 1.

Action (c) Sample Reserve: If the current observation $(\mathbf{z}^t, \mathbf{y}^t)$ does not satisfy the sample deletion criterion or the neuron growing criterion or the parameter update criterion, then the sample is pushed to the rear end of the data stream for a later use. Due to the self-adaptive nature of the thresholds, these reserve samples may also contain some useful information and will be used later in the learning process.

These three actions of self-regulating learning are repeated for all the samples in the training sequence.

CSRAN algorithm is summarised and given below in Pseudocode 5form.

8.5.3 *Guidelines for the Selection of the Self-regulatory Thresholds*

In this section, we explain the influence of the fixed and self-regulatory thresholds on the approximation ability of CSRAN and provide some guidelines for their initialization. It should be noted that the instantaneous sample magnitude error (M_t^e) and the normalized absolute sample phase error (ϕ_t^e) are defined such that their values are in the range $[0,2]$. The self-regulatory learning algorithm works by dividing the sample error region into three different sub-regions, namely the sample deleting region, the network parameters updating region and the neuron addition region as shown in Fig. 8.4.

When the sample magnitude and phase errors fall within the **sample deleting region**, the corresponding sample is deleted from the training sequence without being used in the learning process. The area of this region depends on the expected approximation accuracy and are controlled by the fixed sample deleting magnitude (E_d^M) and phase (E_d^ϕ) thresholds. Increasing E_d^M and E_d^ϕ above the desired accuracy results in deletion of many samples from the training sequence. But, the resultant network may not satisfy the desired performance. Hence, they are fixed at the expected accuracy level.

The basic principle behind the CSRAN algorithm is that the sample with a higher residual error contains significant information and must be used first to add a new hidden neuron. The significant information is identified using the self-regulatory

Pseudocode 5 Pseudo code for CSRAN Algorithm.

```
Input: Present the training data one-by-one to the network
                    from the data stream.
Output: Size of the network, parameters of the network.
START
  For each input (z^t) compute the network output (ŷ^t)
      ŷ^t_l = Σ_{j=1}^h v_lj sech[σ_j^T (z^t − c_j)];  l = 1,··· n;
  Using residual error (e^t = y^t − ŷ^t),
      estimate M_t^e and φ_t^e using Eqs. (8.19) and (8.20).
    Sample Deletion: If M_t^e ≤ E_d^M AND φ_t^e ≤ E_d^φ
      then delete the sample from the sequence without learning.
    Sample Learning: During learning, either a hidden neuron is
          added or pruned or the network parameters are updated.
          Neuron Growing Criterion: If M_t^e ≥ E_a^M OR φ_t^e ≥ E_a^φ,
          then allocate new hidden neuron with v_{h+1} = e_t, c_{h+1} = z_t,
          and σ_{h+1} = κ (z_t − c_nr). Also, update the neuron growing
          thresholds using Eq. (8.24) and
          increase the dimensionality of P_t.
          Parameter Update Criterion: If M_t^e ≥ E_l^M OR φ_t^e ≥ E_l^φ
          then update CSRAN network parameters using
          the C-EKF. Also, update the learning thresholds
          using Eq. (8.27).
          Neuron Pruning Criterion: Identify the non-contributing
          neuron for N_w consecutive samples and remove it.
          The dimensionality of P_t is reduced by removing
          the rows/columns corresponding to the pruned neuron.
     Sample Reserve: When a sample does not satisfy deletion,
     growing and update criteria, it is pushed to the rear end
     of data stream.
     Continue steps 1 to 5 until there are no more samples in the
     training data stream.
END
```

thresholds E_a^M and E_a^ϕ. If these thresholds are chosen closer to the maximum residual error, then very few neurons will be added to the network. Such a network will not approximate the function properly. If a lower value is chosen, the resultant network may contain many neurons with poor generalization ability. Hence, the range for the growing control parameters can be selected in the interval of $50\% - 95\%$ of the maximum residual error. Since the growing thresholds are self-regulatory, they will be decreasing based on the current sample error. Hence, the lower limit on the growing thresholds can be taken as the initial value of the learning thresholds.

Intuitively, one can see that the learning thresholds (E_l^M and E_l^ϕ) can be chosen lower than the growing thresholds. The learning thresholds are initialized at a

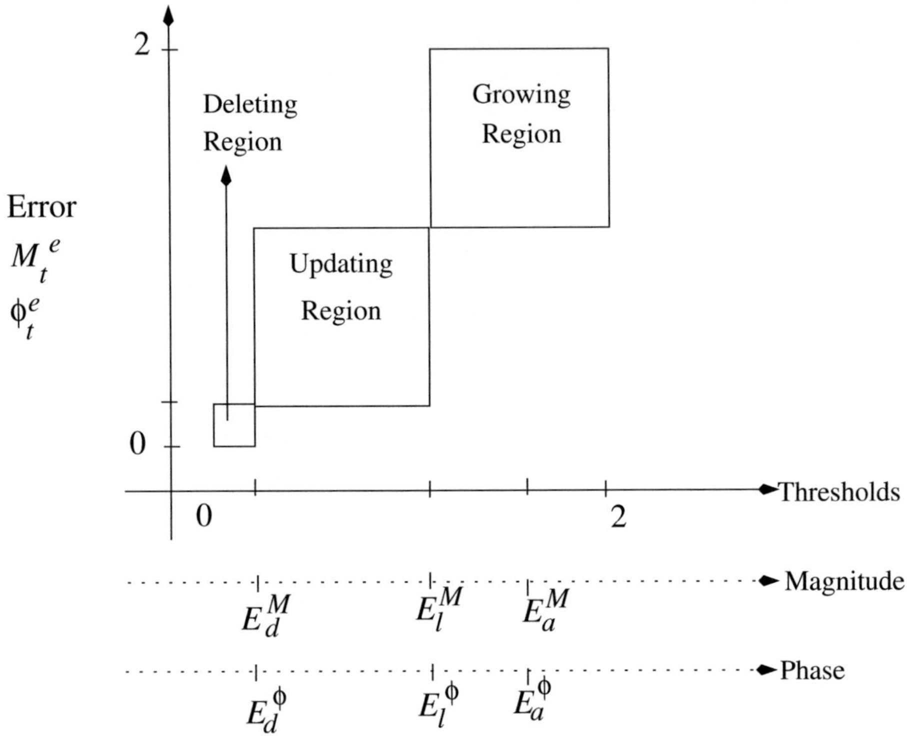

Fig. 8.4 Error Regions for the Various Self-regulatory Thresholds of the CSRAN

value that is at most equal to the lowest value of the growing thresholds after self-regulation, i.e., between 10%-40% of the maximum error values. If these thresholds are chosen closer to 50% of maximum residual error, then very few samples will be used for adapting the network parameters and most of the samples will be pushed to the end of the training sequence. The resultant network will not approximate the function accurately. If a lower value is chosen, then all samples will be used in updating the network parameters with-out altering the training sequence. Learning similar samples repeatedly results in over-training, thereby, affecting the generalization ability of the network. Similar to the growing thresholds, the learning thresholds are also self-regulatory in nature. As the learning thresholds undergo self-regulation, the lowest values these thresholds can reach will be their respective sample deleting control parameters.

The process of self-regulation is controlled by the slope parameter δ. The slope parameter controls the rate at which the self-regulatory thresholds reach its minimum. In general, the slope parameter is initialized close to 1. Larger the slope,

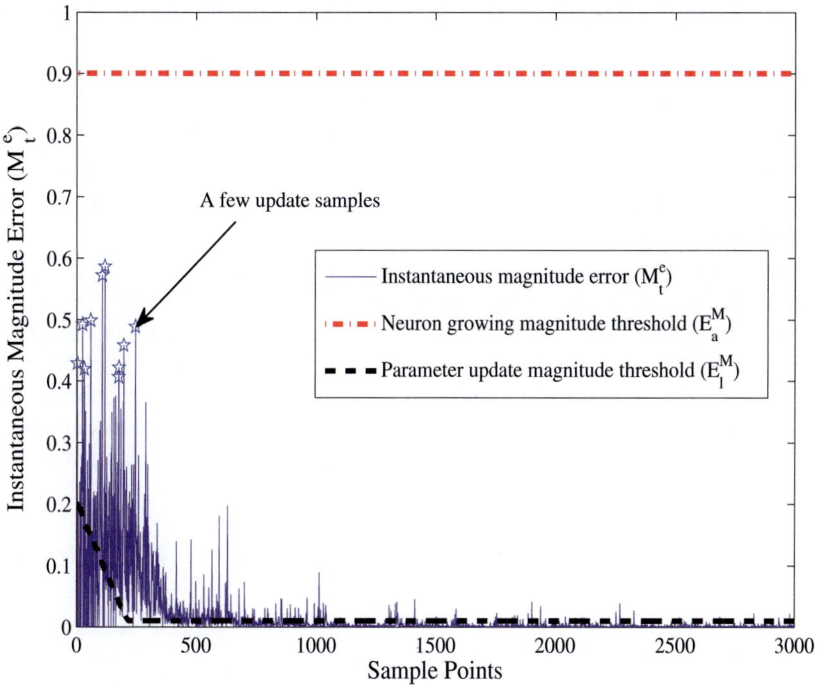

Fig. 8.5 Sample magnitude error and self-regulating magnitude growing and update thresholds history for SCFAP-I.

slower will be the rate at which the neuron growing thresholds will reach their minimum value and hence this will result in fewer neurons being added to the network.

Another important parameter which remove the non-contributing neurons is the neuron pruning control parameter (E_p). Similar to CMRAN, the CSRAN algorithm removes non-contributing neurons for N_w consecutive samples. The only difference is that the CMRAN uses the neuron's contribution in real and imaginary part for deletion, where as the CSRAN uses the contribution in magnitude and phase. If the pruning control parameter (E_p) is set at a higher value, pruning seldom occurs, and all the added neurons will remain in the network irrespective of their contribution to the network output. This might result in a larger network structure. Contrarily, lower value of E_p results in frequent pruning, resulting in oscillations and insufficient neurons to the network.

It is to be noted that, for the CSRAN, all thresholds except the growing (E_a^M and E_a^ϕ) and learning control parameters (E_l^M and E_l^ϕ) are fixed for all problems.

8.5.4 Illustration of the Self-regulatory Learning Principles Using a Complex-valued Function Approximation Problem

Here, we use the Synthetic Complex Function Approximation Problem I (SCFAP-I) given in Chapter 5 to illustrate the key features (especially the self-regulating part) of CSRAN. The thresholds in CSRAN are initialized as: Magnitude thresholds: $\{E_d^M = 0.002,\ E_a^M = 0.9;\ E_l^M = 0.2\}$; Phase thresholds: $\{E_d^\phi = 0.002;\ E_a^\phi = 1.8;\ E_l^\phi = 0.4\}$, and other parameters: $\{\delta = 0.9995;\ E_p = 0.08;\ N_w = 50;\ \kappa = 0.83;\ p_0 = 1.05$ and $q = 0.005\}$.

Fig. 8.5 shows the sample magnitude error (M_t^e), the neuron growing magnitude threshold (E_a^M) and the parameter update magnitude threshold (E_l^M) for the 3000 uniformly sampled training data. From the figure, we can see that the sample magnitude error is always lesser than 0.6 ($< E_a^M$) and hence no neurons are added based on the magnitude error (i.e., no regulation in the neuron growing magnitude threshold). The influence of M_t^e in the adaptation of the network parameters can be seen from the history of the E_l^M (shown as dashed line in Fig.8.5). Few sample points

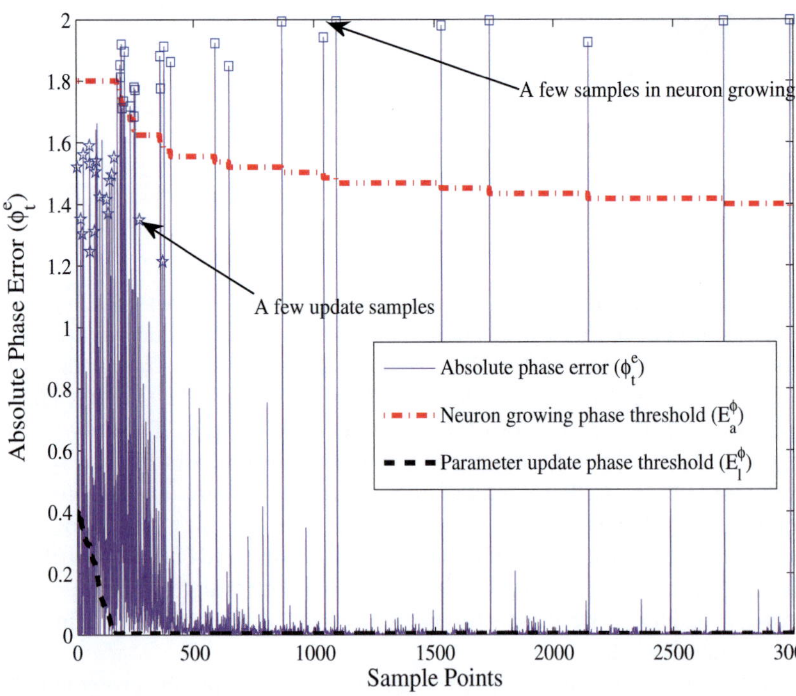

Fig. 8.6 Sample absolute phase error and self-regulating phase growing and update thresholds history for SCFAP-I.

used for the network parameter update are highlighted (using the symbol '$\star$' in Fig. 8.5). Since M_t^e is greater than E_l^M at these points, the samples are used for parameter updates and the E_l^M is adapted. After approximately 300 points, E_l^M reaches a minimum threshold value (E_d^M).

The absolute phase error (ϕ_t^e), the neuron growth (E_a^ϕ) and the parameter update (E_l^ϕ) thresholds are shown in Fig. 8.6. From the figure, we can observe that the ϕ_t^e at some sample instants (highlighted using symbol '$\square$') is greater than E_a^ϕ and hence new hidden neurons are added at these sample instants. Also, the threshold (E_a^ϕ) is adapted based on the error as shown using the 'dash-dot' line. From Figs. 8.5 and 8.6, we can observe that the magnitude and absolute phase errors for the first 179 sample instants are lower than the neuron growing thresholds (E_a^M and E_a^ϕ) and are either used in network parameter update or shifted to the rear-end of the training data stream.

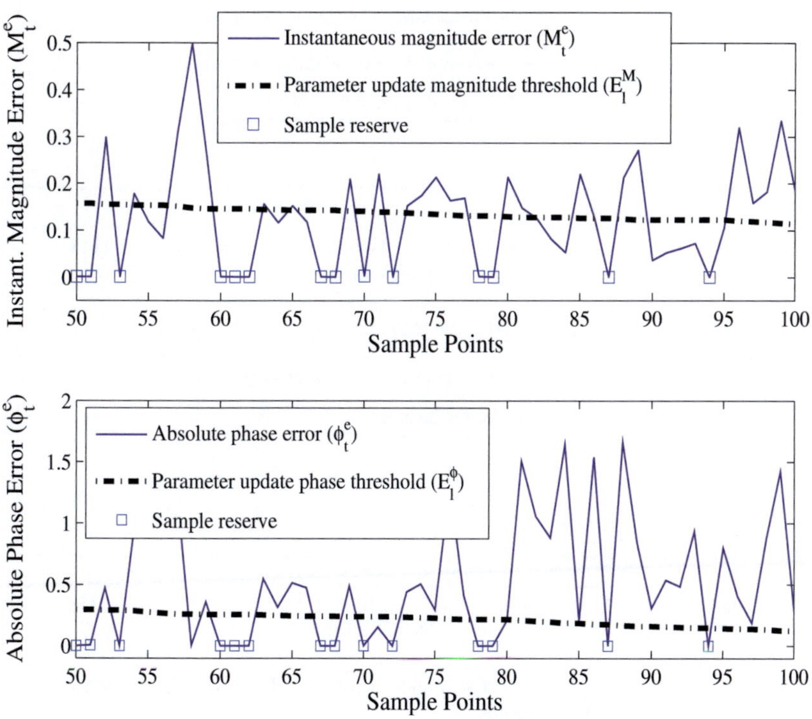

Fig. 8.7 A snapshot of magnitude error, absolute phase error, and self-regulating magnitude/phase update thresholds history between 50 – 100 samples.

In order to clearly highlight the 'sample learning' and the 'sample reserve' mechanisms, we present a snap-shot of the sample instants $50-100$ in Fig. 8.7. From the snap-shot, we can see that the parameter update magnitude and phase thresholds are self-regulating, if and only if M_t^e is greater than E_l^M or ϕ_t^e is greater than E_l^ϕ. Otherwise, the current sample is skipped from learning and is pushed to the rear-end of the training sequence. For example, at sample instants $57-59$, M_t^e is greater than E_l^M and hence these samples are used to update the network parameters. Also, E_l^M is self-adapted using the Eq. (8.27) at these sample instants. At the sample point 90, even though M_t^e is less than E_l^M, the sample is used for a network parameter update as $\phi_t^e > E_l^\phi$. At samples $78-79$ (marked with '□' symbol in Fig. 8.7), both M_t^e and ϕ_t^e are lower than their respective thresholds and are therefore, shifted to the rear end of the training data sequence.

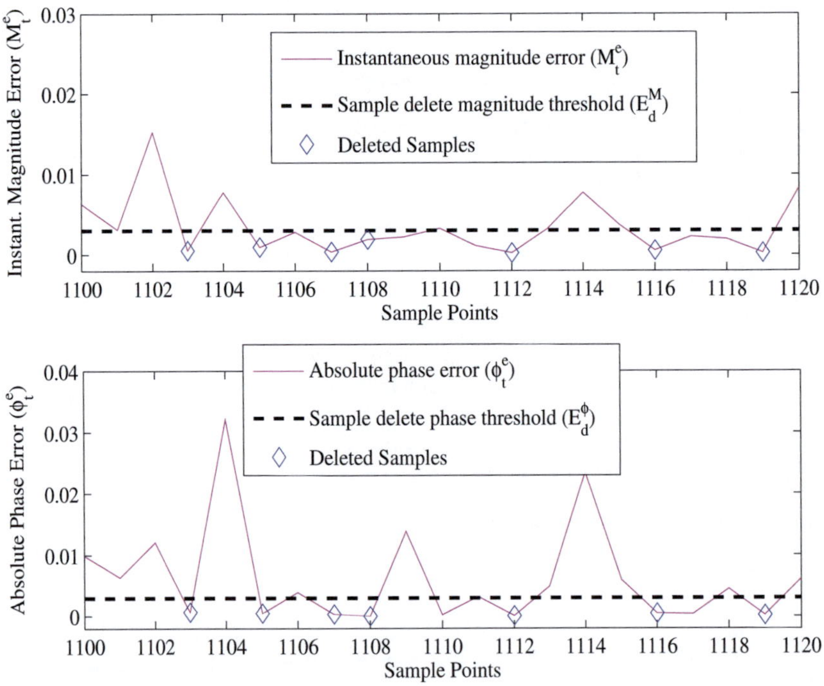

Fig. 8.8 A snapshot of magnitude error, phase error, and CSRAN magnitude/phase delete thresholds between 1100-1120 samples.

Next, we present another snap-shot between sample instants $1100-1120$ in Fig. 8.8 to illustrate the 'sample deletion' mechanism. From Fig. 8.8, we can see that M_t^e and ϕ_t^e are lower than their deletion thresholds at sample points marked with '◇' symbol. These samples are deleted without being used for updating the network parameters. Even though M_t^e in sample points 1106, and 1118 are smaller than E_d^M, they are used in the parameter update as ϕ_t^e at these sample instants are greater than E_d^ϕ.

The neuron history for samples that are used for learning is shown in Fig. 8.9. Out of the 3000 training samples, only 1726 samples are used for the learning process, 230 samples are deleted from the training set and 1044 samples are shifted to the rear-end of the training data stream. Out of these 1044 samples, only 197 samples are used for fine tuning the network parameters and the remaining samples are deleted from the sequence. At the end of the training stage, only 1923 samples are used in the learning process and 1077 samples are deleted from the training sequence.

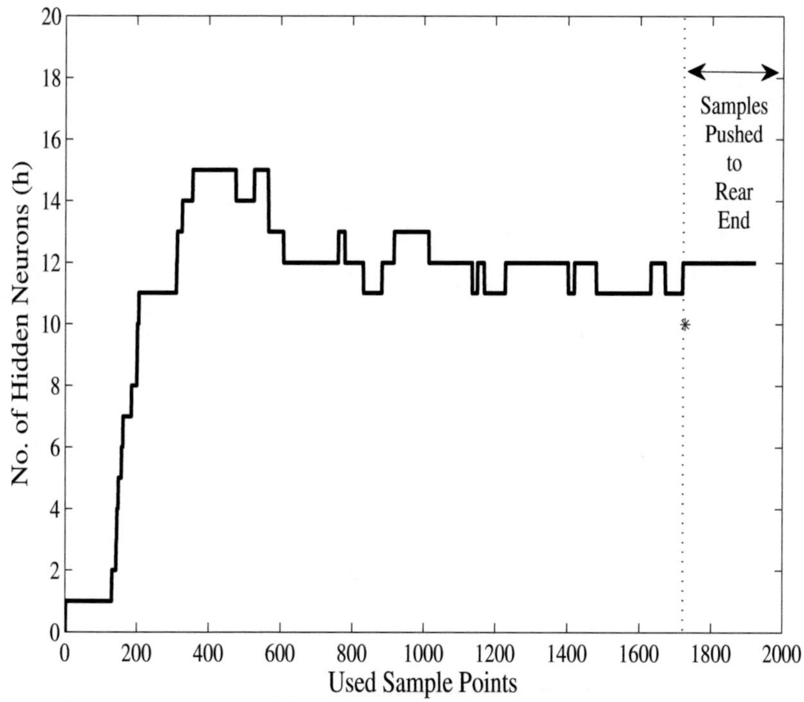

Fig. 8.9 Neuron history for SCFAP-I.

Thus, one can clearly see the sel-regulating mechanism of CSRAN from the above discussion and in the next section we present a detailed performance evaluation of CSRAN.

8.6 Performance Study: Complex-valued Function Approximation Problems

8.6.1 Complex-valued Function Approximation Problem I

First, we present the performance of CSRAN using the CFAP-I defined in the previous section and compare it with existing complex-valued learning algorithm in the literature. The results for CMRAN, FC-MLP, CRBF and FC-RBF are reproduced from [5, 12].

Table 8.1 presents the training and testing performance measures (J_{Me} and Φ_e), number of neurons (h) and training time used for sequential and batch learning algorithms considered in this study. From the table, it can be seen that CSRAN algorithm outperforms the existing sequential learning CMRAN. With only 12 neurons, the magnitude and phase errors are at least 20 times lower than that of CMRAN results. Due to the self-regulating scheme, CSRAN uses only 1923 samples of the total 3000 samples, and hence the computational effort is much lower than that of the CMRAN algorithm.

From the table, one can also note that CSRAN outperforms CRBF and C-ELM, whereas the performance of FC-RBF is similar to that of CSRAN. Since CSRAN is a sequential learning algorithm with a self-regulating scheme, it requires lesser number of neurons and computational effort. Also, it can be inferred from the table that the C-ELM, being analytical, requires the lowest computation time to approximate the function. However, its magnitude and phase errors are significantly higher than that of CSRAN algorithm.

8.6.2 Complex-valued Function Approximation Problem II

We consider another synthetic Complex Function Approximation Problem II (CFAP-II) as defined in Chapter 5. Similar to CFAP-I, training samples are presented for 5000 epochs for the batch learning FC-MLP, CRBF, and FC-RBF algorithms. The best learning rate, initial weights and number of neurons for FC-MLP, CRBF, FC-RBF are chosen as discussed in [12, 5]. The growing and

Table 8.1 Performance Comparison for CFAP-I

Nature of Algorithm	Algorithm	Time (sec.)	h	Training Error		Testing Error	
				J_{Me}	Φ_e (deg.)	J_{Me}	Φ_e (deg.)
Sequential	CSRAN	31	12	0.003	0.7	0.003	0.31
	CMRAN	84	27	0.068	12.6	0.07	15.81
Batch	CRBF	5020	20	0.59	45.3	0.62	47.15
	C-ELM	0.22	20	0.69	34.9	0.704	36.15
	FC-RBF	1341	15	0.0019	0.3	0.003	0.34
	FC-MLP	1423	15	0.005	0.4	0.006	0.61

update thresholds of the proposed CSRAN have been initialized as: Magnitude thresholds: $\{E_a^M = 0.9; E_l^M = 0.2\}$ and Phase thresholds: $\{E_a^\phi = 1.8; E_l^\phi = 0.4\}$.

Figure 8.10 shows the neuron growth history for CSRAN as the samples are learnt sequentially from the training data. From the figure, we can see that 2 hidden neurons are added first and two additional neurons are added from the samples pushed to the rear end of data stream.

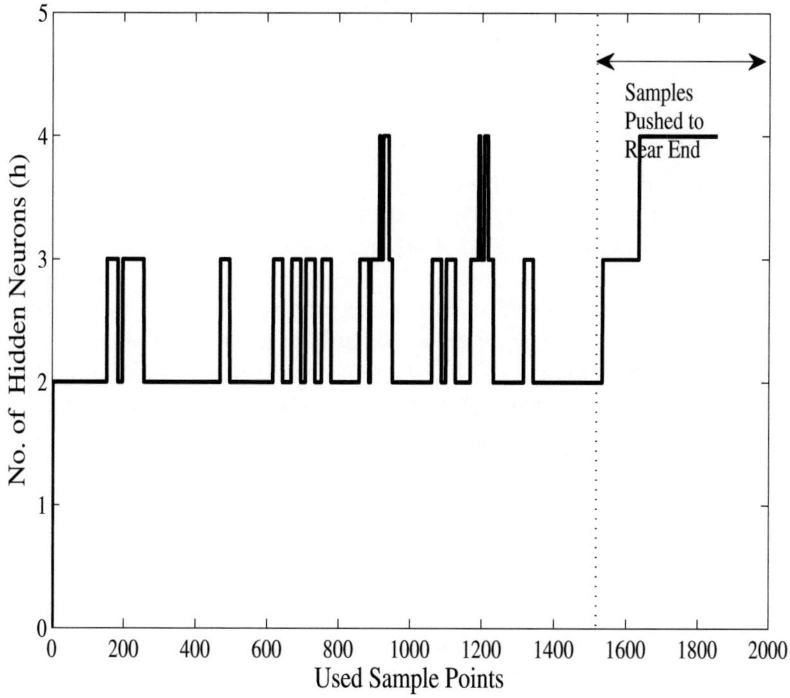

Fig. 8.10 Neuron history for CFAP-II.

Out of the 3000 randomly chosen samples, only 1512 samples are used in the learning process, 34 samples are deleted from the training set and the remaining 1454 samples are shifted to the rear-end of the data stream. The 1454 samples are used again to fine tune the network parameters. Here, only 338 samples are used for fine tuning the network parameters and 2 samples are used for adding new hidden neurons. The rest 1114 are deleted from the training sequence. At the end of the training stage, only 1852 samples are used for training and 1148 samples are deleted from the training sequence.

Table 8.2 gives the quantitative performance comparison of the proposed CSRAN with CMRAN and also other well-known batch learning complex-valued learning algorithms. For CFAP-II, the proposed CSRAN algorithm uses only 1852 samples

Table 8.2 Performance Comparison for CFAP-II

Nature of Algorithm	Algo.	Time (sec)	h	Training		Testing	
				J_{Me}	Φ_e	J_{Me}	Φ_e
Sequential	CSRAN	18	4	0.008	2.2	0.02	2.8
	CMRAN	52	14	0.0257	2.2	0.5	19
Batch	CRBF	9686	15	0.1465	51.2	0.2	52
	C-ELM	0.2	15	0.1917	90	0.2	88.2
	FC-RBF	1909	20	0.0196	15.9	0.05	15.8
	FC-MLP	1857	15	0.029	15.7	0.05	15.6

out of the 3000 training samples. Here CSRAN requires only 4 neurons to approximate the function, whereas the existing CMRAN requires 14 neurons to approximate the function. The training and generalization performances of CSRAN clearly indicate that it outperforms the CMRAN and the other well-known batch learning algorithms.

Next, we compare the performance of the proposed CSRAN algorithm with the well known batch learning algorithms available in literature, viz., the C-ELM, the I-ELM, the FC-RBF and the FC-MLP. Comparing the performance of CSRAN with that of the other batch learning algorithms, in general, the average absolute phase error of CSRAN algorithm is at least 5 times smaller than the other well-known algorithms. Next, we evaluate the performance of CSRAN on two real world problems.

8.6.3 QAM Channel Equalization Problem

The performance of CSRAN equalizer is evaluated in comparison to the various complex-valued neural equalizers are evaluated using the magnitude error (J_{Me}), the average absolute phase error (Φ_e) and the symbol-error-rate (SER). The thresholds of CSRAN are initialized as: $\{E_a^M = 1.8,\ E_a^\phi = 1.8\}$; $\{E_l^M = 0.2,\ E_l^\phi = 0.2\}$; and the other parameters are same as in CFAP-I. For the batch learning networks used in the study, the training sample set is presented 1000 times repeatedly. Finally, the results of all these algorithms are also compared with the Bayesian equalizer, which is optimal (the best result one can achieve) [2].

From the results, we find that there are no reserve samples. Out of 5000 samples, 4710 samples are used for learning and the remaining 290 samples are deleted. For QAM problem, CSRAN used 7 neurons to capture the nonlinear relationship between the channel output and the transmitted symbol.

Table 8.3 summarizes the training/testing performances of the different complex-valued neural equalizers, the number of hidden neurons (h) used and the training time. From the table, it can be seen that J_{Me} and Φ_e of CSRAN is lower than all the other equalizers chosen for the comparison, except that I-ELM has lower magnitude error than CSRAN. However, I-ELM requires 100 neurons to achieve this magnitude

Table 8.3 Performance Comparison for the Nonlinear Non-Minimum Complex Phase Equalization Problem at 20dB

Nature of	Algorithm	Time (sec)	h	Training		Testing	
				J_{Me}	Φ_e (deg.)	J_{Me}	Φ_e (deg.)
Sequential	CSRAN	45	7	0.2	6.4	0.38	7.09
	CMRAN	555	40	0.4	17.47	0.43	21.17
Batch	CRBF	8107	15	0.6	35.19	0.6	39.86
	C-ELM	0.36	15	0.6	34.14	0.58	35.11
	FC-RBF	3840	15	0.4	31.17	0.41	14.69
	FC-MLP	3862	15	0.2	6.47	0.72	31.1

error, whereas CSRAN requires 7 neurons to obtain a comparable magnitude error and a minimum phase error. Fig. 8.11 gives the SER plot for the different equalizers. It can be observed that CSRAN has a lower SER at a SNR of 20dB. However, at lower SNR, the effects of all the equalizers are almost similar. CSRAN equalizer achieves an SER performance of -3 (approximately) at 17dB SNR whereas the optimal Bayesian equalizer achieves the same SER performance at 15dB. Other complex-valued neural equalizers attain the same performance at a higher SNR.

8.6.4 Adaptive Beam Forming Problem

The results for the CRBF, C-ELM, FC-RBF and FC-MLP beam-formers are reproduced from [12]. The self-regulating thresholds of CSRAN for the adaptive beam-forming problem are initialized as: $\{E_a^M = 1.0,\ E_a^\phi = 1.5\}$ and $\{E_l^M = 0.1,\ E_l^\phi = 0.1\}$. The other parameters are same as in CFAP-I.

CMRAN [13] uses 32 neurons to perform beam-forming while CSRAN requires only 6 neurons to perform beam-forming. Moreover, CSRAN beam-former uses only 234 samples in the learning process. The gains for the signals and interference nulls for the various beam-formers are summarized in Table 8.4. From the table, it can be observed that CSRAN beam-former outperforms all the other complex-valued beam-formers and its performance is also slightly better than the conventional optimal matrix method [14].

As the orthogonal decision boundaries of a complex-valued network makes them well suited for solving real-valued classification problems. Hence, we also study the classification performance of CSRAN using real-valued benchmark problems from the UCI machine learning repository [15].

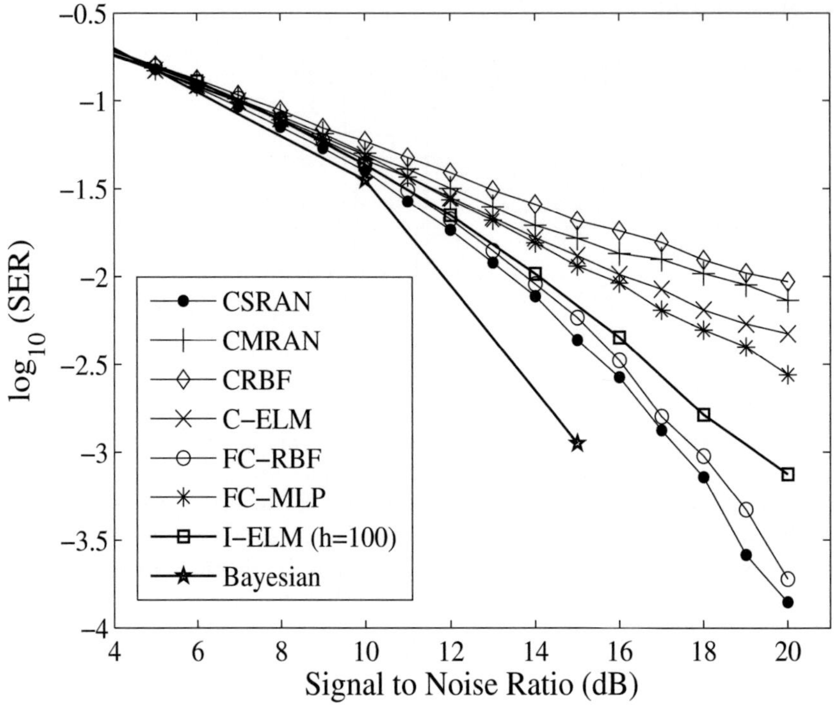

Fig. 8.11 Error Probability Curve for Various Complex-valued Neural Equalizers

Table 8.4 Performance Comparison of Various Complex-valued Neural Network Based Beam-formers

DOA[a]	Gain (dB)						
	CSRAN	CMRAN	CRBF	C-ELM	FC-RBF	FC-MLP	MM[b]
Beam-1: -30°	-13.83	-16.84	-17.94	-18.05	-16.99	-13.98	-13.98
Null-1: -15°	-59.56	-49.77	-27.53	-48.28	-58.45	-53.99	-57.02
Null-2: 0°	-57.37	-49.6	-27	-41.64	-57.23	-53.99	-57
Null-3: 15°	-57.93	-48.18	-28.33	-47.5	-56.32	-53.99	-57.02
Beam-2: 30°	-13.93	-17.52	-17.92	-16.68	-17.00	-13.98	-13.98

[a] Direction of Arrival (DoA)
[b] Matrix Method (MM)

Table 8.5 Performance comparison for the multi-category classification problems.

Problem	Classifier Domain	Learning Model	No. of neurons	Training time (Sec.)	Training η		Testing η	
					η_o	η_a	η_o	η_a
Image Segmentation	Real valued	SVM	127	721	94.28	94.28	91.38	91.38
		ELM	49	0.25	92.86	92.86	90.23	90.23
		SRAN	48	22	97.62	97.62	93	93
	Complex valued	FC-RBF	38	421	96.19	96.19	92.33	92.33
		BB-CELM	65	0.3	96.19	96.19	92.5	92.5
		CSRAN	54	339	92	92	88	88
Vehicle Classification	Real-valued	SVM	340	550	79.48	79.82	70.62	68.51
		ELM	150	0.4	85.14	85.09	77.01	77.59
		SRAN	113	55	91.45	90.25	75.12	76.86
	Complex valued	FC-RBF	70	678	88.67	88.88	77.01	77.46
		BB-CELM	100	0.11	90.33	90.16	80.3	80.4
		CSRAN	80	352	84.19	84.24	79.15	79.16
Glass Identification	Real-valued	SVM	183	320	86.24	93.23	70.47	75.61
		ELM	80	0.05	92.66	96.34	81.31	87.43
		SRAN	59	28	94.49	96.28	86.2	80.95
	Complex valued	FC-RBF	90	452	95.6	95.54	83.76	80.95
		BB-CELM	70	0.08	93.58	82.92	88.16	81
		CSRAN	80	452	87.16	93.67	83.5	78.09

8.7 Performance Study: Real-valued ClassificationProblems

Next, the performance of CSRAN is studied on 3 multi-category and 5 binary benchmark real-valued classification problems. In all these problems, the performance of CSRAN is compared with the complex-valued FC-RBF, and BB-CELM classifiers. In addition, performances are also compared with the real-valued SVM, ELM, and SRAN classifiers. For the classification problems, the various self-regulating thresholds of CSRAN are initialized as: $\{E_a^M = 1.6, E_a^\phi = 1.0\}$; $\{E_l^M = 0.6, E_l^\phi = 0.6\}$.

The performance results of CSRAN classifier on the three benchmark multi-category classification problems are presented in Table 8.5. It can be observed from the table that the performance of CSRAN classifier is comparable to those of the other classifiers in all the three problems. It must be noted here that CSRAN has been originally developed to solve function approximation problems, while SRAN, FC-RBF and BB-CELM have been developed to solve classification problems. However, the performance of CSRAN can be further improved by developing the classifier using the hinge loss error function defined in Eq. (6.16).

Next, the performance of CSRAN on the four binary benchmark classification problems is presented in Table 8.6. From the results in the table, it can be observed that the performance of CSRAN is almost similar to those of the other complex-valued and best performing real-valued classifiers considered in this study.

Table 8.6 Performance comparison on benchmark binary classification problems.

Problem	Classifier Domain	Classifier	K	Training Time (s)	Training Efficiency		Testing Efficiency	
					(η_o)	(η_a)	(η_o)	(η_a)
Liver disorders	Real-valued	SVM	141	0.0972	79.5	77.93	71.03	70.21[5]
		ELM	100	0.1685	88.5	87.72	72.41	71.41
		SRAN	91	3.38	92.5	91.8	66.9	65.8
	Complex-valued	FC-RBF	20	133	77.25	75.86	74.46	75.41
		BB-CELM	15	0.06	77.5	75.69	75.17	74.64
		CSRAN	20	38	71	72.41	67.59	69.24
PIMA data	Real-valued	SVM	221	0.205	77	74.71	77.45	76.33
		ELM	100	0.2942	84.3	82.64	76.63	75.25
		SRAN	97	12.24	89	87.35	78.53	74.9
	Complex-valued	FC-RBF	20	130.3	72	65.77	78.53	68.49
		BB-CELM	10	0.15	76.5	74.77	78.8	77.31
		CSRAN	20	64	75.5	73.13	77.99	76.73
Breast cancer	Real-valued	SVM	24	0.1118	98.67	98.76	96.6	97.06
		ELM	66	0.1442	100	100	96.35	96.5
		SRAN	7	0.17	98	97.5	96.87	97.3
	Complex-valued	FC-RBF	10	158.3	99	99.02	97.12	97.45
		BB-CELM	15	0.06	94.33	94.39	92.69	91.78
		CSRAN	20	60	98.67	98.57	96.08	96.86
Iono-sphere	Real-valued	SVM	43	0.0218	97	95.83	91.24	88.51
		ELM	32	0.0396	94	92.27	89.64	87.52
		SRAN	21	3.7	99	98.6	90.84	91.88
	Complex-valued	FC-RBF	10	186.2	98	97.83	89.64	88.01
		BB-CELM	25	0.17	95	94.27	88.85	85.67
		CSRAN	3	86	94	91.67	88.05	85.78
Heart Disease	Real-valued	SVM	42	0.038	87.14	86.69	75.5	75.1
		ELM	36	0.15	90	89.58	76.5	75.9
		SRAN	28	0.534	91.43	90.83	78.5	77.525
	Complex-valued	FC-RBF	20	45.6	94.29	94.17	78	77.78
		BB-CELM	5	0.03	84.3	82.5	83	82.53
		CSRAN	20	26	100	100	76.5	76.21

Hence, it can be inferred from the study that the performance of CSRAN is almost similar to other complex-valued and real-valued classifiers used in the study.

8.8 Summary

To summarize, a sequential learning algorithm with a self-regulating scheme has been developed in this chapter for a complex-valued resource allocation network.

The self-regulating scheme uses both the explicit magnitude and phase thresholds to control the 'sample deletion', 'sample learning' and 'sample reserve' mechanisms. The sample learning includes growing/pruning the hidden neurons and updating the network parameters using a complex-valued extended Kalman filter algorithm. The self-regulating scheme in CSRAN helps in achieving better generalization performance with a compact network structure. A complex-valued synthetic function approximation has been used to clearly describe the various key features of CSRAN algorithm. The performance of CSRAN has been evaluated using two synthetic complex-valued function approximation problems, a complex-valued QAM channel equalization problem and an adaptive beam-forming problem. The results indicate the superior performance of CSRAN algorithm. From a set of real-valued benchmark classification problems, we have shown the decision making ability of the proposed CSRAN.

References

1. Deng, J.P., Sundararajan, N., Saratchandran, P.: Communication channel equalization using complex-valued minimal radial basis function neural networks. IEEE Transactions on Neural Networks 13(3), 687–696 (2002)
2. Li, M.B., Huang, G.B., Saratchandran, P., Sundararajan, N.: Complex-valued growing and pruning RBF neural networks for communication channel equalisation. IEE Proceedings- Vision, Image and Signal Processing 153(4), 411–418 (2006)
3. Yingwei, L., Sundararajan, N., Saratchandran, P.: A sequential learning scheme for function approximation using minimal radial basis function neural networks. Neural Computation 9(2), 461–478 (1997)
4. Huang, G.B., Saratchandran, P., Sundararajan, N.: An efficient sequential learning algorithm for growing and pruning RBF (GAP-RBF) networks. IEEE Transactions on Systems, Man, and Cybernetics - Part B: Cybernetics 34(6), 2284–2292 (2004)
5. Savitha, R., Suresh, S., Sundararajan, N., Saratchandran, P.: A new learning algorithm with logarithmic performance index for complex-valued neural networks. Neurocomputing 72(16-18), 3771–3781 (2009)
6. Goh, S.L., Mandic, D.P.: An augmented extended kalman filter algorithm for complex-valued recurrent neural networks. Neural Computation 19(4), 1039–1055 (2007)
7. Cha, I., Kassam, S.A.: Channel equalization using adaptive complex radial basis function networks. IEEE Journal on Selected Areas in Communications 13(1), 122–131 (1995)
8. Suksmono, A.B., Hirose, A.: Intelligent beamforming by using a complex-valued neural network. Journal of Intelligent and Fuzzy Systems 15(3-4), 139–147 (2004)
9. Amin, M.F., Islam, M.M., Murase, K.: Ensemble of single-layered complex-valued neural networks for classification tasks. Neurocomputing 72(10-12), 2227–2234 (2009)
10. Huang, G.B., Li, M.B., Chen, L., Siew, C.K.: Incremental extreme learning machine with fully complex hidden nodes. Neurocomputing 71(4-6), 576–583 (2008)
11. Huang, G.B., Chen, L., Siew, C.K.: Universal approximation using incremental constructive feedforward networks with random hidden nodes. IEEE Transactions on Neural Networks 17(4), 879–892 (2006)
12. Savitha, R., Suresh, S., Sundararajan, N.: A fully complex-valued radial basis function network and its learning algorithm. International Journal of Neural Systems 19(4), 253–267 (2009)

13. Jianping, D., Sundararajan, N., Saratchandran, P.: Complex-valued minimal resource allocation network for nonlinear signal processing. International Journal of Neural Systems 10(2), 95–106 (2000)
14. Monzingo, R.A., Miller, T.W.: Introduction to Adaptive Arrays. SciTech. Publishing, Raleigh (2004)
15. Blake, C., Merz, C.: UCI repository of machine learning databases. Department of Information and Computer Sciences. University of California, Irvine (1998), `http://archive.ics.uci.edu/ml/`

MIX
Papier aus verantwortungsvollen Quellen
Paper from responsible sources
FSC® C105338

If you have any concerns about our products,
you can contact us on
ProductSafety@springernature.com

In case Publisher is established outside the EU,
the EU authorized representative is:
Springer Nature Customer Service Center GmbH
Europaplatz 3, 69115 Heidelberg, Germany

Printed by Libri Plureos GmbH
in Hamburg, Germany